MACMILLAN/McGRAW-HILL

Math

Daily Practice Workbook
with Summer Skills
Refresher

Grade 5

McGraw Hill

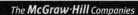

Contents

Daily Practice

Summer Skills Refresher

Name _____

Benchmark Numbers

Use a benchmark number to help you decide which is the more reasonable number.

1. If the container at left holds 100 paper clips, about how many paper clips can the container at right hold: 50 or 500?

2. If the box at left has 500 rubber bands, about how many rubber bands does the box at right have: 2,000 or 20,000?

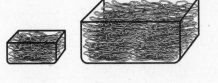

_____

3. If the jar at left has 150 buttons, about how many buttons does the jar at right have: 450 or 1,500?

4. If the photo album at left holds 125 photos, about how many photos does the photo album at right hold: 250 or 500?

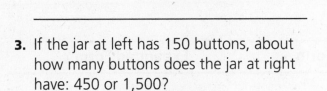

_____

5. If the crate at left holds about 50 oranges, about how many oranges does the crate at right hold: 100 or 1,000?

6. If the machine at left has 200 toy animals, about how many toy animals does the machine at right have: 400 or 1,000?

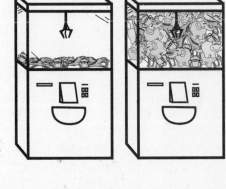

_____

Name _____

Place Value Through Billions

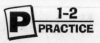
Name the place and value of each underlined digit.

1. 2,3<u>4</u>6 _____

2. 6<u>5</u>,893 _____

3. 7<u>6</u>3,406,594 _____

4. 40<u>7</u>,356,138,920 _____

5. 64,<u>3</u>21,008 _____

6. 1<u>1</u>7,927,724,417 _____

7. 903,00<u>4</u>,200,006 _____

Complete the table.

	Standard Form	Short Word Form	Expanded Form
8.		3 thousand, 125	
9.		52 thousand, 40	
10.			7,000,000 + 400,000 + 50,000 + 600 + 90 + 3
11.		200 million, 80 thousand, 9	
12.			80,000,000,000 + 1,000,000,000 + 200,000,000 + 40,000,000 + 5,000,000 + 800 + 70
13.	9,000,000,006		
14.		452 billion, 370 million	

Problem Solving
Solve.

15. Mercury is the planet closest to the Sun. It orbits the Sun from a distance of about 28 million, 600 thousand miles. Write the number in standard form.

_____

16. Pluto is the planet farthest from the Sun. It orbits the Sun from a distance of about 4 billion, 551 million, 400 thousand miles. Write the number in standard form.

_____

Use with Grade 5, Chapter 1, Lesson 2, pages 4–7.

Name _____

Explore Decimal Place Value

Use the 10-by-10 grids to model each decimal.

1. 0.37

2. 0.01

3. 1.62

Use 10-by-10 grids to help you complete the table.

	Decimal	Number of Ones	Number of Tenths	Number of Hundredths
4.	5.09			
5.	3.74			
6.	4.81			
7.	0.57			

Use with Grade 5, Chapter 1, Lesson 3, pages 8–9.

Decimal Place Value

Write the place value of the underlined digit.

1. 7.8̲1 _____

2. 0.12̲6 _____

3. 14.600̲5 _____

4. 20.9̲03 _____

5. 5.07̲8 _____

6. 0.50̲94 _____

7. 9.15̲76 _____

8. 7.681̲5 _____

Complete the table.

	Standard Form	Short Word Name	Expanded Form
9.	4.16		
10.		13 and 78 thousandths	
11.			0.9 + 0.03
12.		20 and 7 thousandths	
13.	0.1392		
14.		2 and 608 ten-thousandths	

Write an equivalent decimal for each.

15. 0.3 _____

16. 1.400 _____

17. 5.00 _____

18. 3.10 _____

19. 2.7 _____

20. 4.75 _____

21. 16.53 _____

22. 9.5100 _____

23. 87.05 _____

Problem Solving

Solve.

24. The fifth grade students at Highland School are cleaning up 3 and 45 hundredths miles of Sugar Creek Forest. Write the number of miles in standard form.

_____

25. In all, the fifth graders picked up 30 and 58 thousandths kilograms of trash. Write the number of kilograms in standard form.

_____

Use with Grade 5, Chapter 1, Lesson 4, pages 10–13.

Compare and Order Whole Numbers and Decimals

Compare. Write >, <, or =.

1. 3,976 ◯ 4,007

2. 89,001 ◯ 89,100

3. 126,698 ◯ 126,689

4. 1,435,052 ◯ 145,052

5. 19,463,674 ◯ 29,436,764

6. 4,303,259,087 ◯ 4,033,259,807

7. 328,574,000,256 ◯ 328,574,010,256

8. 2.7 ◯ 2.82

9. 6.030 ◯ 6.03

10. 7.89 ◯ 7.189

11. 12.54 ◯ 1.254

12. 0.981 ◯ 2.3

13. 0.004 ◯ 0.040

Order from least to greatest.

14. 17,639; 3,828; 45,947 _____

15. 890,409; 890,904; 809,904 _____

16. 21,997; 29,979; 219,997; 21,797 _____

17. 5,630,168; 5,036,168; 6,530,168; 563,168 _____

18. 8.26; 8.02; 8.6 _____

19. 58.50; 5.085; 5.85; 5.805 _____

20. 0.186; 0.1; 0.86; 0.168 _____

21. 5.309; 5.003; 0.53; 0.9 _____

Problem Solving
Solve.

22. In January, the average low temperature in Montreal, Quebec, Canada, is 5.2°F, and the average low temperature in Cape Town, South Africa, is 60.3°F. Which city is warmer in January?

23. In one year Seattle, Washington, recorded 0.24 inches of snow, Chicago, Illinois, recorded 30.9 inches of snow, and Birmingham, Alabama, recorded 1 inch of snow. Write these amounts in order from least to greatest.

_____

_____

Name _____

Problem Solving: Skill
Use the Four-Step Process

Solve. Use the four-step process.

1. The three highest mountains in Colorado are Mount Massive (14,421 ft), Mount Harvard (14,420 ft), and Mount Elbert (14,433 ft). Which mountain has the greatest height?

_____

2. Hoover Dam, in the United States, is 223 meters high. Ertan Dam, in China, is 240 meters high. In Canada, Mica Dam is 243 meters high. List the dams by height from greatest to least.

_____

_____

3. The Akshi Kaikyo suspension bridge in Japan has a span of 6,570 feet. The Humber suspension bridge in England has a span of 4,626 feet. The Izmit Bay suspension bridge in Turkey has a span of 5,538 feet. Which bridge has the shortest span?

_____

4. There are three long tunnels that go under Boston Harbor. The Sumner Tunnel is 5,653 feet long. The Callahan Tunnel is 5,070 feet long. The Ted Williams Tunnel is 8,448 feet long. List the tunnels from shortest to longest.

_____

_____

Mixed Strategy Review

Use the data from the table for problems 5–7.

5. Which tunnel is the longest?

_____

6. List the tunnels by name in order from shortest to greatest.

_____

_____

7. Which tunnels are fewer than 5,000 feet long?

_____

Land Tunnels in the United States		
Tunnel	**State**	**Length (ft)**
Liberty Tubes	Pennsylvania	5,920
Devil's Side	California	3,400
E. Johnson Memorial	Colorado	8,959
Squirrel Hill	Pennsylvania	4,225

Use with Grade 5, Chapter 1, Lesson 6, pages 18–19.

Name _____

Add and Subtract
Whole Numbers and Decimals

Add or subtract.

1.	9,868 + 6,329	2.	3,136 − 473	3.	0.87 + 6.12	4.	4.45 − 1.02
5.	3,007 − 1,980	6.	4.672 + 15.31	7.	31,043 + 56,691	8.	2.85 − 0.58
9.	4.609 − 2.81	10.	124,543 + 96,883	11.	12.974 + 4.734	12.	20,431 − 17,642
13.	5.8 + 4.289	14.	30,048 − 9,338	15.	$1.09 − 0.65	16.	76,509 + 120,306
17.	321,658,400 − 197,369,250	18.	3,472,196 + 7,810,984	19.	3.65 − 0.824	20.	$28.99 + 1.75

21. 34,504 + 5,712 = _____   22. 1.265 + 8.77 = _____

23. 9.54 − 4.883 = _____   24. 2,980 + 135,618 = _____

25. $44.65 − $2.19 = _____   26. 78,327 − 59,912 = _____

27. $0.33 + $5.79 = _____   28. 210,336 − 89,481 = _____

Problem Solving
Solve.

29. Gasoline prices are given to the nearest thousandth of a dollar. If gasoline rises in price from $1.499 to $1.589, what is the amount of the increase?

30. The area of Texas is 695,676 square kilometers. That is 525,368 more square kilometers than Florida. What is the area of Florida in square kilometers?

_____   _____

Name _____

Estimate Sums and Differences

Round to the underlined place.

1. 2,741 _____

2. 8.37 _____

3. $315.95 _____

4. 34,098 _____

5. 79.437 _____

6. 58.164 _____

Round to the place indicated.

7. 653 (ten) _____

8. 2.468 (hundredth) _____

9. $105.49 (dollar) _____

10. 39,281.7 (hundred) _____

11. 46.275 (tenth) _____

12. $1.6195 (cent) _____

Estimate each sum or difference. Show your work.

13.   317
   + 288

14.   0.88
   + 6.336

15.   1,642
   − 381

16.   3.09
   − 2.98

17. 6.461 − 3.1085

_____

18. 38 + 504 + 81

_____

19. 5,319 − 1,999

_____

20. $13.77 + $9.95

_____

21. 3,498 − 734

_____

22. $76.08 − $61.97

_____

Problem Solving
Solve.

23. The driving distance from Los Angeles to Chicago is 2,054 miles. The distance from Chicago to Boston is 983 miles. About how many miles is the drive from Los Angeles to Boston through Chicago?

_____

24. Rita paid for a $2.95 sandwich and a $1.19 drink with a $10 bill. About how much did her lunch cost? About how much did she receive in change?

_____

Use with Grade 5, Chapter 2, Lesson 2, pages 28–31.

Problem Solving: Strategy
Find a Pattern

Find a pattern to solve.

1. A student just learning the high jump starts with the bar at 3 feet. The pole is raised 0.4 inch after each successful jump. How high will the bar be after 5 successful jumps?

   _____

2. A beginning pole vaulter raises the bar 0.5 inch after each successful vault. On the first jump the bar is at 4 feet 5 inches. How high will the bar be after 3 successful attempts?

   _____

3. **Art** A designer is making a tile mosaic. The first row of the mosaic has 1 red tile in the center. If the designer increases the number of red tiles in the center of each row by 4, how many red tiles will be in the center of the fifth row?

   _____

4. **Health** Brian has started an exercise program in which he walks daily. He plans to increase the distance that he walks by 0.25 mile each week. He walks 2.25 miles everyday the first week. How many miles will he be walking each day during the fifth week?

   _____

Mixed Strategy Review

Solve. Use any strategy.

5. **Number Sense** The sum of two whole numbers between 20 and 40 is 58. The difference of the two numbers is 12. What are the two numbers?

   _____

   Strategy: _____

6. Ramon has $3.50. He buys two pens that cost $0.75 each and a pencil that costs $0.40. How much money does Ramon have left?

   _____

   Strategy: _____

7. Denise earns $25 delivering papers each week. She saves $2.00 the first week. She plans to increase her savings by $3.00 each week, until she is saving $20.00 every week. In how many weeks will she have her first $20 in savings?

   _____

   Strategy: _____

8. **Create a problem** for which you could find a pattern to solve. Share it with others.

   _____

   _____

   _____

Properties of Addition

Identify the addition property used to rewrite each problem.

1. 59 + 83 = 83 + 59

_____

2. 0 + 426 = 426

_____

3. (33 + 42) + 17 = 33 + (42 + 17)

_____

4. 66 + 27 + 24 = 66 + 24 + 27

_____

5. 3.09 + 0 = 3.09

_____

6. 3.1 + (7.2 + 0.6) = (3.1 + 7.2) + 0.6

_____

Add or subtract. Describe your work.

7. 3 + 9 + 7 = _____

8. 14 + 11 + 56 = _____

9. 97 + 74 = _____

10. 76 − 18 = _____

11. $2.25 + $5.75 = _____

12. 12 + 194 + 88 = _____

13. 568 − 29 = _____

14. 2.41 + 3.6 = _____

15. 249 + 98 = _____

16. 6 + 3 + 2 + 4 + 7 = _____

17. $234 − $96 = _____

18. 27 + 42 = _____

19. 196 − 21 = _____

20. $3.60 + $5.40 = _____

21. 1 + 53 + 9 + 7 = _____

22. 9 + (127 + 13) = _____

23. 624 − 302 = _____

24. 3.78 + 1.04 = _____

25. 396 + 504 = _____

26. 3.72 − 1.97 = _____

27. 247 + (47 + 53) = _____

28. (0.092 + 0.008) − 0.1 = _____

Problem Solving
Solve.

29. Brandon's lunch order totaled $3.94. He gave the cashier $10.00. How much money should he get back?

_____

30. To get home from school, Kara walks 4 minutes to the bus, rides the bus for 28 minutes, and walks 6 minutes to her house. How long is her trip?

_____

Use with Grade 5, Chapter 2, Lesson 4, pages 34–35.

Name _____

Mental Math: Addition and Subtraction

Add or subtract. Describe your work.

1. 43 + 108 = _____

2. 79 − 16 = _____

3. 63 − 28 = _____

4. 58 + 97 = _____

5. 774 + 238 = _____

6. 451 + 391 = _____

7. 635 − 469 = _____

8. 785 − 294 = _____

9. 531 + 279 = _____

10. 641 − 462 = _____

11. 8.2 + 7.5 = _____

12. 14.3 − 8.4 = _____

13. 21.3 − 9.8 = _____

14. 41.6 + 62.7 = _____

15. 3.18 + 4.97 = _____

16. 6.09 + 7.84 _____

17. 9.64 − 3.95 = _____

18. 0.73 − 0.46 = _____

19. 59.87 + 102.9 = _____

20. 74.63 − 34.8 = _____

Problem Solving
Solve.

21. Maria spent $9.50 on a movie ticket and $2.79 on a drink. How much money did Maria spend in all?

_____

22. The Millburn Cinema has two theaters. Theater One has 477 seats. Theater Two has 296 seats. How many more seats are in Theater One than in Theater Two?

_____

23. A shirt has a regular price of $37.29. It is on sale for $19.79. How much do you save by buying the shirt on sale?

_____

24. Jed saves $62 in June, $78 in July, and $92 in August. How much does he save in all?

_____

Name _____

Choose the Method: Addition and Subtraction

When you add and subtract, you can use pencil and paper, a calculator, or mental math.

Add or subtract. Tell which method you use.

1. 309 + 556 = _____

2. 264 − 73 = _____

3. 1,715 + 662 = _____

4. 8,105 − 586 = _____

5. 3,643 − 2,210 = _____

6. 6,217 + 9,843 = _____

7. 2,709 + 3,427 = _____

8. 9,001 − 3,224 = _____

9. $61.04 + $8.97 = _____

10. 13.78 − 4.32 = _____

11. 3.624 − 0.462 = _____

12. 14.17 + 29.34 = _____

13. $425.08 − $379.58 = _____

14. 142.96 + 75.8 = _____

15. $123.18 + $721.76 = _____

16. 509.2 − 328.9 = _____

Algebra Find each missing number. Describe your work.

17. 2.16 + _____ = 3.67

18. 320 − _____ = 46.8

19. 1,206 + _____ = 2,571

20. 48.72 − _____ = 23.42

Problem Solving
Solve.

21. Erin sells a necklace for $17.99 and a bracelet for $14.99. How much money does she receive in all?

_____

22. A stadium has 46,495 seats. If 27,818 people come to watch a game, how many seats are empty?

_____

Use with Grade 5, Chapter 2, Lesson 6, pages 38–39.

Patterns of Multiplication

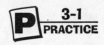

Complete.

1. 8 × 2 = _____

8 × 20 = _____

8 × 200 = _____

8 × 2,000 = _____

2. 6 × 4 = _____

6 × 40 = _____

6 × 400 = _____

6 × 4,000 = _____

3. 4 × 5 = _____

4 × 50 = _____

4 × 500 = _____

4 × 5,000 = _____

4. 3 × 80 = _____

30 × 80 = _____

300 × 80 = _____

3,000 × 80 = _____

5. 5 × 60 = _____

50 × 60 = _____

500 × 60 = _____

5,000 × 60 = _____

6. 9 × $70 = _____

90 × $70 = _____

900 × $70 = _____

9,000 × $70 = _____

7. 4 × $11 = _____

40 × $11 = _____

400 × $11 = _____

4,000 × $11 = _____

8. 7 × 8 = _____

70 × 8 = _____

700 × 8 = _____

7,000 × 8 = _____

9. 2 × $5 = _____

20 × $5 = _____

200 × $5 = _____

2,000 × $5 = _____

Multiply.

10. 90 × 3 = _____

11. 7 × $4,000 = _____

12. 200 × 6 = _____

13. 30 × 40 = _____

14. 600 × 70 = _____

15. 40 × 800 = _____

16. 4 × $1,000 = _____

17. 500 × 80 = _____

18. 70 × 100 = _____

19. 3 × 30 = _____

20. 5 × 1,000 = _____

21. 7 × $900 = _____

22. 50 × 80 = _____

23. 100 × 80 = _____

24. 50 × 20 = _____

Problem Solving

Solve.

25. The 9 members of a music club in Indianapolis want to fly to New York to see several musicals. The cost of a round trip ticket is $300. How much would the airfare be altogether?

_____

26. During one week, an airport shop sold 70 New York City travel guides for $9 each. How much was the total received for the guides?

_____

Name _____

Explore the Distributive Property

Multiply.

1. 7 × 19 = _____

2. 6 × 22 = _____

3. 8 × 58 = _____

4. 5 × 13 = _____

5. 4 × 76 = _____

6. 2 × 27 = _____

7. 9 × 56 = _____

8. 3 × 71 = _____

9. 7 × 33 = _____

10. 8 × 34 = _____

11. 4 × 83 = _____

12. 3 × 27 = _____

13. 6 × 88 = _____

14. 9 × 98 = _____

15. 5 × 65 = _____

16. 5 × 36 = _____

17. 3 × 98 = _____

18. 2 × 97 = _____

Rewrite each problem using the Distributive Property.

19. 3 × 13

_____

20. 8 × 68

_____

21. 7 × 32

_____

22. 9 × 35

_____

23. 8 × 17

_____

24. 4 × 71

_____

25. 5 × 25

_____

26. 6 × 84

_____

Problem Solving
Solve.

27. Each of 6 hikers were allowed to bring 24 pounds of gear on a cross-country hike. How many pounds of gear was that altogether?

_____

28. The hikers plan to travel an average of 12 miles each day for 9 days. How many miles do they plan to travel in all?

_____

Use with Grade 5, Chapter 3, Lesson 2, pages 56–57.

Multiply Whole Numbers

Multiply.

1. $3 \times 5,012 =$ _____

2. $7 \times 2,436 =$ _____

3. $4 \times 12,261 =$ _____

4. $43 \times \$65 =$ _____

5. $458 \times 26 =$ _____

6. $329 \times 72 =$ _____

7. $58 \times 1,036 =$ _____

8. $94 \times 5,425 =$ _____

9. $33 \times 24,918 =$ _____

10. $328 \times 142 =$ _____

11. $179 \times 212 =$ _____

12. $826 \times \$507 =$ _____

13. 371 $\times \quad 4$	**14.** \$507 $\times \quad 7$	**15.** 7,693 $\times \quad 8$	**16.** 29,148 $\times \quad 3$	**17.** 345 $\times \quad 42$

18. \$740 $\times \quad 16$	**19.** 3,006 $\times \quad 28$	**20.** 26,308 $\times \quad 25$	**21.** 449 $\times \; 515$	**22.** 762 $\times \; 108$

Compare. Write $>$, $<$, or $=$.

23. $63 \times 25 \bigcirc 31 \times 78$

24. $5 \times 5,026 \bigcirc 52 \times 189$

25. $5 \times 123 \bigcirc 15 \times 41$

26. $835 \times 95 \bigcirc 83 \times 803$

27. $47 \times 6,351 \bigcirc 64 \times 11,382$

28. $597 \times 13 \bigcirc 24 \times 806$

29. $48 \times 212 \bigcirc 5,227 \times 4,968$

30. $43 \times 321 \bigcirc 98 \times 65$

31. $4 \times 49 \times 7 \bigcirc 9 \times 65 \times 2$

32. $12 \times 58 \times 29 \bigcirc 37 \times 37 \times 42$

Problem Solving

Solve.

33. A basketball player scored an average of 23 points per game. He played 82 games during the season. How many points did he score that season?

_____

34. A basketball arena has 36 sections of seats. Each section contains 784 seats. How many people can the arena seat?

_____

Properties of Multiplication

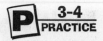

Identify the multiplication property used to rewrite each problem.

1. $49 \times 0 = 0$

_____

2. $3 \times 8 = 8 \times 3$

_____

3. $1 \times 67 = 67$

_____

4. $7 \times (56 - 3) = (7 \times 56) - (7 \times 3)$

_____

5. $2 \times (36 + 93) = (2 \times 36) + (2 \times 93)$

_____

6. $2 \times (9 \times 8) = (2 \times 9) \times 8$

_____

7. $1.41 \times 12 = 12 \times 1.41$

_____

8. $0 \times 5.4 = 0$

_____

9. $(7 \times 5) \times 8 = 7 \times (5 \times 8)$

_____

10. $74 \times 1 = 74$

_____

Multiply. Name the property you used.

11. $7 \times 28 =$ _____

12. $5 \times 25 =$ _____

13. $0 \times 96 =$ _____

14. $1 \times 36 =$ _____

15. $8 \times 72 =$ _____

16. $6 \times 34 =$ _____

17. $4 \times 53 =$ _____

18. $11 \times 11 =$ _____

19. $50 \times 102 =$ _____

Fill in the number that makes each sentence true.

20. $6 \times (2 + 8) = (6 \times$ ____$) + (6 \times$ ____$)$

21. ____ $\times 7.8 = 7.8$

22. $0 \times 65 =$ ____

23. $(2.1 \times 0.9) \times 8.6 = 2.1 \times ($ ____ $\times 8.6)$

24. $7.6 \times 6.4 = 6.4 \times$ ____

25. ____ $\times 1 = 54$

26. $53 \times (18 \times$ ____$) = (53 \times 18) \times 9$

27. $75 \times$ ____ $= 83 \times 75$

Problem Solving
Solve.

28. Tony displayed his model race cars in 3 rows of 11 cars each. How else could Tony have displayed his cars in equal rows?

_____

29. Sarah displayed her stuffed animals on 2 shelves. Each shelf contains 2 rows of 9 animals. How many animals does Sarah have?

_____

Use with Grade 5, Chapter 3, Lesson 4, pages 62–65.

Name _____

Estimate Products of Whole Numbers and Decimals

Estimate by rounding.

1. 3.4 × 10 _____   **2.** 59 × 32 _____   **3.** 446 × 682 _____

4. 816 × 1.04 _____   **5.** 4.27 × 82 _____   **6.** 83 × 303 _____

7. 21 × 663 _____   **8.** 98 × 32 _____   **9.** 91 × 3.2 _____

10. 3.34 × 847 _____   **11.** 9.29 × 0.8 _____   **12.** 43 × 58 _____

13. 8.9 × 4.5 _____   **14.** 13.1 × 0.6 _____   **15.** 87.2 × 65.8 _____

16. 186 × 92 _____   **17.** 342 × 86 _____   **18.** 396 × 23 _____

19. 631 × 465 _____   **20.** 0.863 × 89.24 _____   **21.** 605 × 7.235 _____

22. 85.47   **23.** 603   **24.** 408   **25.** 3,045   **26.** 6.34
  × 83.6   × 29   × 46   × 38   × 6

27. 0.8   **28.** 27.43   **29.** 8.5   **30.** 5.13   **31.** 3,498
  × 5.2   × 8   × 38   × 24   × 5.7

Estimate by clustering.

32. 236 + 186 + 209 _____   **33.** 42.8 + 36.9 + 41.9 _____

34. 5,497 + 4,623 + 4,802 _____   **35.** 9.07 + 8.7 + 9.45 _____

36. 739 + 662 + 720 _____   **37.** 11.4 + 9.68 + 10.5 _____

Problem Solving
Solve.

38. Mia bought 2.5 lb of sliced turkey to make sandwiches for a picnic. The turkey cost $5.89 per pound. About how much did Mia pay for the turkey?

39. Mia also bought a package of sliced cheese that weighed 2.38 lb. The cheese cost $4.25 per pound. About how much did Mia pay for the cheese?

_____   _____

Use with Grade 5, Chapter 3, Lesson 5, pages 66–69.

17

Name _____

Problem Solving: Skill
Estimate or Exact Answer

Solve. Then state whether the problem calls for an estimate
or an exact answer.

1. The ski team has a race at 9:00 A.M. The race is 120 miles away. The team leaves at 6:00 A.M. and drives about 50 miles each hour. Will they arrive at the race on time?

_____

2. The ski team travels in 4 vans. Each van holds 9 team members. How many members are on the team?

_____

3. School raffle tickets cost $8 apiece. The school's goal is to raise at least $3,000 from the raffle. If 424 tickets are sold, will the school meet its goal?

_____

4. Students at Tuscan School filled out a survey. The survey showed that of 374 students, 195 speak a second language. How many students speak only one language?

_____

Mixed Strategy Review

5. Book World receives 12 boxes of books. Each box contains 16 copies of the new best-seller, *Norton's Last Laugh*. How many copies of *Norton's Last Laugh* does the store receive?

_____

_____

6. At the beginning of the last year, there were 368 students at the elementary school. By the beginning of this year, 72 of those students had moved. Were there more or less than 300 return students? How do you know?

_____

_____

7. A brochure says that Yellowstone National Park covers 2,220,000 acres. Do you think that this number is exact or an estimate? Explain your answer.

_____

_____

8. Hunter School has kindergarten and grades 1–6. There are 2 kindergarten classes and 2 classes in each grade. If the maximum class size is 25, what is the greatest number of students that could be in the school?

_____

_____

Use with Grade 5, Chapter 3, Lesson 6, pages 70–71.

Multiply Whole Numbers by Decimals

Multiply.

1. 1.6
$\times$ 8

2. 2.83
$\times$ 7

3. 14.7
$\times$ 24

4. 3.75
$\times$ 100

5. 2.09
$\times$ 8

6. 12.8
$\times$ 10

7. 2.55
$\times$ 42

8. 4.7
$\times$ 85

9. $34.99
$\times$ 4

10. 147.4
$\times$ 2

11. $0.8 \times 5 =$ _____

12. $1.67 \times 4 =$ _____

13. $6 \times \$1.79 =$ _____

14. $2.46 \times 10 =$ _____

15. $4.2 \times 22 =$ _____

16. $10.4 \times 1,000 =$ _____

17. $2.3 \times 38 =$ _____

18. $57 \times 5.18 =$ _____

Find the multiple of 10 that makes each statement true.

19. $6.1 \times$ _____ $= 610$

20. _____ $\times 11.84 = 118.4$

21. $\$24.95 \times$ _____ $= \$249.50$

22. $526.7 \times$ _____ $= 526,700$

23. $0.2687 \times$ _____ $= 268.7$

24. $0.46 \times$ _____ $= 46$

25. _____ $\times 32.05 = 3,205$

26. $0.012 \times$ _____ $= 0.12$

Problem Solving

Solve.

27. Each Sunday during his nine-week summer vacation, Ray buys a newspaper. The Sunday paper costs $1.85. How much did Ray spend on the Sunday newspaper during his vacation?

28. One Sunday, Ray weighed the newspaper. It weighed 2.7 lb. If each Sunday newspaper weighs the same, how many pounds of newspaper will Ray recycle if he buys the Sunday paper for 50 weeks?

_____

_____

Name _____

Explore Multiplying Decimals by Decimals

Multiply using the 10-by-10 grids.

1. $0.4 \times 0.7 =$ _____

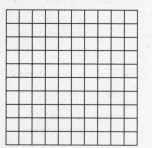

2. $0.7 \times 0.3 =$ _____

3. $0.2 \times 0.8 =$ _____

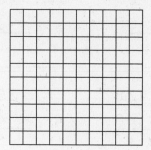

4. $0.4 \times 0.5 =$ _____

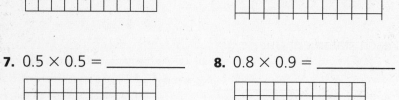

5. $0.6 \times 0.5 =$ _____

6. $0.3 \times 0.2 =$ _____

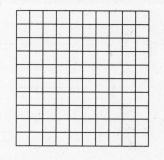

7. $0.5 \times 0.5 =$ _____

8. $0.8 \times 0.9 =$ _____

9. $0.1 \times 0.8 =$ _____

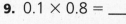

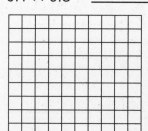

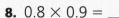

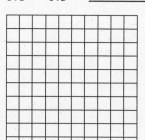

Problem Solving
Solve.

10. Van bought a poster for his room that measures 0.6 m by 0.4 m. Shade the grid to find the area of the glass Van needs to cover the poster. What is the area of the glass?

_____

Multiply Decimals by Decimals

Multiply.

1. $\begin{array}{r} 0.6 \\ \times\ 0.8 \\ \hline \end{array}$	**2.** $\begin{array}{r} 0.5 \\ \times\ 0.6 \\ \hline \end{array}$	**3.** $\begin{array}{r} 1.7 \\ \times\ 0.9 \\ \hline \end{array}$	**4.** $\begin{array}{r} 2.61 \\ \times\ 0.4 \\ \hline \end{array}$	**5.** $\begin{array}{r} 2.09 \\ \times\ 0.3 \\ \hline \end{array}$
6. $\begin{array}{r} 5.18 \\ \times\ 2.7 \\ \hline \end{array}$	**7.** $\begin{array}{r} 6.09 \\ \times\ 8.6 \\ \hline \end{array}$	**8.** $\begin{array}{r} 37.24 \\ \times\ 3.1 \\ \hline \end{array}$	**9.** $\begin{array}{r} 218.7 \\ \times\ 4.8 \\ \hline \end{array}$	**10.** $\begin{array}{r} 432.1 \\ \times\ 1.2 \\ \hline \end{array}$

11. $0.9 \times 0.7 =$ _____

12. $0.16 \times 0.6 =$ _____

13. $7.4 \times 0.4 =$ _____

14. $3.47 \times 0.9 =$ _____

15. $4.35 \times 1.7 =$ _____

16. $58.2 \times 6.8 =$ _____

17. $3.06 \times 9.1 =$ _____

18. $94.2 \times 2.5 =$ _____

19. $17.64 \times 3.2 =$ _____

20. $41.38 \times 6.3 =$ _____

21. $86.51 \times 0.8 =$ _____

22. $0.53 \times 9.7 =$ _____

Find the number that makes each problem true.

23. $\begin{array}{r} 3\ 9.\ 8 \\ \times\ \quad 0.\ 7 \\ \hline 2\ 7.\ \quad 6 \end{array}$ ☐

24. $\begin{array}{r} 4\ 6.\ 8\ 7 \\ \times\ \qquad 0.\ 5 \\ \hline 2\ 3.\ \quad 3\ 5 \end{array}$ ☐

25. $\begin{array}{r} 2.\ 3 \\ \times\ 1.\ 8 \\ \hline .\ 1\ 4 \end{array}$ ☐

26. $\begin{array}{r} 5\ 7.\ 8 \\ \times\ \quad 0.\ 7 \\ \hline 4\ \quad .\ 4\ 6 \end{array}$ ☐

Problem Solving
Solve.

27. Beth works as a lifeguard at a city park. She earns $9.50 per hour and works 7.5 hours each day. How much does she earn each day?

_____

28. The cost of renting a pedal boat at the city park is $6.25 per hour. Jason rented a boat for 1.5 hours. To the nearest cent, how much did the pedal boat rental cost?

_____

Use with Grade 5, Chapter 4, Lesson 3, pages 82–85.

Name _____

Choose the Method: Multiplication

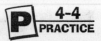

When you multiply, you can use pencil and paper, a calculator, or mental math.

Use the Distributive Property to write an equivalent problem. Then solve.

1. 8 × 699 _____

2. 41 × 43 _____

3. 58 × 52 _____

4. 302 × 9 _____

5. 96 × 18 _____

6. 37 × 62 _____

7. 5.1 × 7.2 _____

8. 2.9 × 90 _____

9. 4.2 × 65 _____

10. 75 × 1.8 _____

Multiply. Tell which method you use.

11. 77 × 97 _____

12. 0.49 × 4 _____

13. 0.9 × 0.08 _____

14. 914 × 35 _____

15. 7.9 × 6.3 _____

16. $1.25 × 7 _____

17. 96 × 99 _____

18. 5 × 4.14 _____

Problem Solving
Solve.

19. Mark earns $12.25 an hour. How much did he make last week if he worked 25 hours?

_____

20. Dan buys a book of 20 stamps. Each stamp costs $0.37. How much does the book of stamps cost?

_____

Use with Grade 5, Chapter 4, Lesson 4, pages 86–87.

Name _____

Problem Solving: Strategy
Guess and Check

Use the guess-and-check strategy to solve.

1. **Science** The Bactrian camel has two humps and the Dromedary camel has one hump. In a group of 15 camels, the total number of humps is 21. How many camels of each type are there?

   _____

2. The circus orders bicycles and unicycles for a new act. It orders a total of 12 cycles. The cycles have 16 tires altogether. How many bicycles and unicycles did the circus order?

   _____

3. Anja buys a magazine and a pizza. She spends $8.10. The magazine costs $2.40 less than the pizza. How much does the pizza cost?

   _____

4. **Social Studies** A letter to Europe from the United States costs $0.80 to mail. A letter mailed within the United States cost $0.37. Nancy mails 5 letters for $2.71, some to Europe and some to the United States. How many letters did she send to Europe?

   _____

Mixed Strategy Review
Solve. Use any strategy.

5. Warren spent $8.50 at the store. He spent $2.40 on paper, $0.88 on pencils, and $2.65 on markers. He spent the rest on a notebook. How much did the notebook cost?

   _____

   Strategy: _____

6. Ms. Baxter takes a group of 8 children to a concert. Tickets for children 12 years and older cost $3.50. Tickets for children under 12 cost $2.25. She spends a total of $21.75 on tickets for the children. How many children are 12 and older?

   _____

   Strategy: _____

7. A cabin has room for 8 campers and 2 counselors. How many cabins are needed for a total of 64 campers and 16 counselors?

   _____

   Strategy: _____

8. **Create a problem** for which you could guess and check to solve. Share it with others.

   _____

   _____

   _____

Exponents

Complete the table.

	Exponent Form	Expanded Form	Standard Form
1.	4^5		
2.		6×6	
3.	1^7		
4.	5^0		
5.		$6 \times 6 \times 6 \times 6$	
6.	10^2		
7.		$2 \times 2 \times 2 \times 2 \times 2 \times 2$	
8.		$3 \times 3 \times 3$	
9.	8^2		
10.	4^1		
11.		$5 \times 5 \times 5 \times 5$	
12.		$10 \times 10 \times 10 \times 10 \times 10 \times 10$	
13.		$7 \times 7 \times 7 \times 7$	
14.	2^3		
15.	9^1		
16.	3^0		

Problem Solving

Solve.

17. There are 10 boxes of postcards. Each box contains 10 bundles of 10 postcards. How many postcards are there altogether? How do you write this number in exponent form?

_____

18. A school has a telephone system for letting families know about emergency school closings. The system is a pyramid with 5 layers. Three parents are in the first layer of the pyramid. Each parent in each layer calls three different parents. How many parents are in the chain?

_____

Use with Grade 5, Chapter 4, Lesson 6, pages 90–93.

Name _____

Relate Multiplication and Division

Divide.

1. 21 ÷ 3 = _____   **2.** 56 ÷ 7 = _____   **3.** 50 ÷ 5 = _____   **4.** 36 ÷ 6 = _____

5. 10 ÷ 2 = _____   **6.** 25 ÷ 5 = _____   **7.** 14 ÷ 2 = _____   **8.** 28 ÷ 4 = _____

9. 72 ÷ 9 = _____   **10.** 18 ÷ 3 = _____   **11.** 48 ÷ 8 = _____   **12.** 36 ÷ 6 = _____

13. 35 ÷ 5 = _____   **14.** 45 ÷ 9 = _____   **15.** 70 ÷ 10 = _____   **16.** 48 ÷ 12 = _____

17. 8)$\overline{64}$ **18.** 4)$\overline{32}$ **19.** 9)$\overline{90}$ **20.** 9)$\overline{36}$ **21.** 11)$\overline{99}$

22. 3)$\overline{24}$ **23.** 5)$\overline{30}$ **24.** 2)$\overline{18}$ **25.** 9)$\overline{99}$ **26.** 7)$\overline{63}$

27. 6)$\overline{42}$ **28.** 8)$\overline{40}$ **29.** 4)$\overline{24}$ **30.** 8)$\overline{32}$ **31.** 6)$\overline{60}$

32. 9)$\overline{81}$ **33.** 5)$\overline{40}$ **34.** 10)$\overline{50}$ **35.** 7)$\overline{49}$ **36.** 9)$\overline{45}$

Complete the fact family.

37. 4 × 3 = 12

3 × 4 = 12

_____ ÷ 3 = 4

_____ ÷ 4 = 3

38. 11 × 8 = 88

8 × 11 = 88

88 ÷ _____ = 11

88 ÷ _____ = 8

39. 9 × 6 = 54

6 × 9 = 54

54 ÷ _____ = 9

54 ÷ 9 = _____

Problem Solving
Solve.

40. Corey has saved 63 files on the hard drive of his computer. He wants to divide them equally among 9 folders. How many files will go in each folder?

41. Jasmine has 48 computer disks. She has just enough cases to place 8 disks in each case. How many cases does she have?

_____   _____

Use Mental Math and Estimate Quotients

 **5-2**
PRACTICE

Complete the pattern.

1. $21 \div 7 = 3$

$210 \div 7 =$ _____

$2,100 \div 7 =$ _____

$21,000 \div 7 =$ _____

2. $48 \div 6 =$ _____

$480 \div 6 =$ _____

_____ $\div 6 = 800$

$48,000 \div 6 =$ _____

3. $36 \div 4 =$ _____

$360 \div 4 =$ _____

$3,600 \div 4 =$ _____

$36,000 \div 4 =$ _____

Divide.

4. $360 \div 6 =$ _____

5. $45,000 \div 5 =$ _____

6. $8,000 \div 100 =$ _____

7. $400,000 \div 8 =$ _____

8. $180 \div 30 =$ _____

9. $16,000 \div 400 =$ _____

10. $49,000 \div 7 =$ _____

11. $5,400 \div 60 =$ _____

12. $72,000 \div 80 =$ _____

Estimate. Use compatible numbers.

13. $231 \div 6$ _____

14. $149 \div 4$ _____

15. $4,748 \div 7$ _____

16. $275 \div 4$ _____

17. $314 \div 6$ _____

18. $5,603 \div 9$ _____

19. $8\overline{)629}$ _____

20. $9\overline{)290}$ _____

21. $9\overline{)342}$ _____

22. $5\overline{)9,461}$ _____

23. $8\overline{)2,943}$ _____

24. $7\overline{)33,875}$ _____

Problem Solving

Solve.

25. Each of the 9 parking lots at an automobile plant holds the same number of new cars. The lots are full. If there are 4,131 cars in the lots, about how many cars are in each lot?

_____

26. A total of 176 valves were used for 8 cars as they were being assembled. About how many valves were used for each car?

_____

Use with Grade 5, Chapter 5, Lesson 2, pages 110–113.

© Macmillan/McGraw-Hill. All rights reserved.

Explore Dividing by 1-Digit Divisors

Draw place-value models to help you divide.

1. 3)463

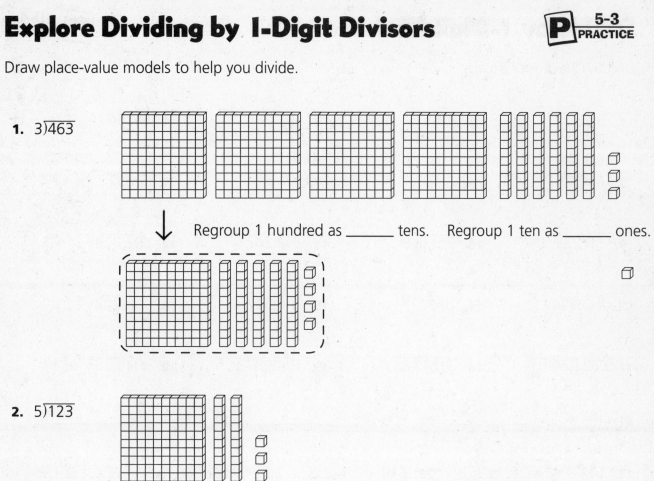

Regroup 1 hundred as _____ tens.   Regroup 1 ten as _____ ones.

2. 5)123

3. 7)226

Name _____

Divide by 1-Digit Divisors

Divide. Check your answer.

1. $3\overline{)385}$ 2. $7\overline{)511}$ 3. $9\overline{)179}$ 4. $5\overline{)254}$

5. $6\overline{)407}$ 6. $8\overline{)167}$ 7. $4\overline{)131}$ 8. $9\overline{)852}$

9. $5\overline{)1,238}$ 10. $3\overline{)3,049}$ 11. $7\overline{)2,242}$ 12. $9\overline{)7,828}$

13. $4\overline{)2,994}$ 14. $8\overline{)2,077}$ 15. $5\overline{)6,322}$ 16. $3\overline{)8,202}$

17. $5\overline{)21,863}$ 18. $3\overline{)74,458}$ 19. $9\overline{)45,373}$ 20. $7\overline{)45,383}$

21. $463 \div 5 =$ _____   22. $606 \div 8 =$ _____   23. $615 \div 2 =$ _____

24. $103 \div 9 =$ _____   25. $618 \div 3 =$ _____   26. $968 \div 6 =$ _____

27. $1,853 \div 2 =$ _____   28. $5,515 \div 4 =$ _____   29. $3,327 \div 8 =$ _____

30. $1,982 \div 3 =$ _____   31. $2,291 \div 9 =$ _____   32. $3,544 \div 5 =$ _____

33. $57,718 \div 8 =$ _____   34. $40,125 \div 4 =$ _____   35. $32,991 \div 6 =$ _____

Problem Soving
Solve.

36. The driving distance between Dallas, Texas, and New York City is 1,604 miles. You make the drive in 4 days and drive the same number of miles each day. How many miles do you drive each day?

_____

37. The distance from New York to Dallas and back is 1,604 miles. You drive from New York to Dallas to New York in July. Your car gets 25 miles for each gallon of gas it uses. To the nearest gallon, how many gallons did you use?

_____

Use with Grade 5, Chapter 5, Lesson 4, pages 116–117.

Name _____

Problem Solving: Skill
Interpreting the Remainder

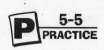

Solve. Tell how you interpreted the remainder.

1. At the Science Fair, a hall featuring an electronics exhibit holds 30 people. How many groups will need to be formed if 460 people want to see the electronics exhibit?

_____

_____

2. Students from four local schools are bussed to the Science Fair. The number of students and teachers attending is 290. Each bus holds 50 passengers. How many buses will be needed?

_____

_____

3. The Natural History Museum is open 10 hours each day. A 9-minute movie about dinosaurs plays continuously. How many times does the complete movie play each day? (Hint: 10 hours = 600 minutes)

_____

_____

4. The Natural History Museum sells postcards for $3 each. Morris has $29 to spend. If he buys as many postcards as he can, how much money will he have left?

_____

_____

5. A museum curator has 203 wildlife photographs. She wants to display them in groups of 8. How many groups of 8 can she make?

_____

_____

6. Each museum tour can have a maximum of 20 people. There are 110 students and teachers who want to take a tour. How many groups will they need?

_____

_____

7. An artist makes models of dinosaurs. He has 59 dinosaurs. He packages them in boxes of 4. After the artist fills as many boxes as he can, how many dinosaurs will be left?

_____

_____

8. At the Discovery Center, students work in groups of 5 or fewer. There are 89 students who want to use the center. What is the least number of groups that will need to be formed?

_____

_____

Name _____

Estimating Quotients Using Mental Math

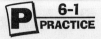

Estimate each quotient. Use compatible numbers.

1. 249 ÷ 81 _____

2. 565 ÷ 94 _____

3. 389 ÷ 53 _____

4. 653 ÷ 69 _____

5. 5,209 ÷ 84 _____

6. 4,116 ÷ 57 _____

7. 1,677 ÷ 39 _____

8. 3,574 ÷ 869 _____

9. 21,076 ÷ 72 _____

10. 13,108 ÷ 433 _____

11. 35,904 ÷ 57 _____

12. 23,096 ÷ 371 _____

13. 161,086 ÷ 179 _____

14. 898,124 ÷ 296 _____

15. $49\overline{)296}$ _____

16. $88\overline{)261}$ _____

17. $53\overline{)4,597}$ _____

18. $621\overline{)1,865}$ _____

19. $196\overline{)3,988}$ _____

20. $78\overline{)6,507}$ _____

21. $27\overline{)59,508}$ _____

22. $394\overline{)314,278}$ _____

23. $537\overline{)409,010}$ _____

24. $62\overline{)474,918}$ _____

25. $924\overline{)823,567}$ _____

26. $72\overline{)135,916}$ _____

Problem Solving

Solve.

27. There are 18,845 fans at a hockey arena. The arena has 24 exits. If the same number of fans uses each exit, about how many fans use each exit?

28. In 1968, Jerry bought $729 worth of stock. In 1998, Jerry sold this stock for $34,096. About how many times the 1968 value was the 1998 value?

_____

_____

Use with Grade 5, Chapter 6, Lesson 1, pages 124–125.

Name _____

Divide by 2-Digit Divisors

Divide. Check your answer.

1. $58\overline{)1,394}$

2. $78\overline{)9,161}$

3. $23\overline{)1,491}$

4. $47\overline{)2,539}$

5. $21\overline{)3,390}$

6. $96\overline{)3,694}$

7. $73\overline{)5,521}$

8. $88\overline{)2,755}$

9. $39\overline{)29,388}$

10. $37\overline{)12,002}$

11. $54\overline{)29,254}$

12. $82\overline{)18,215}$

13. $84\overline{)416,275}$

14. $22\overline{)189,416}$

15. $32\overline{)224,418}$

16. $1,204 \div 33 =$ _____

17. $9,649 \div 84 =$ _____

18. $61,129 \div 95 =$ _____

19. $6,720 \div 45 =$ _____

20. $49,201 \div 70 =$ _____

21. $6,639 \div 87 =$ _____

22. $45,488 \div 96 =$ _____

23. $36,289 \div 54 =$ _____

24. $349,205 \div 75 =$ _____

25. $296,878 \div 42 =$ _____

Problem Solving
Solve.

26. Members of the Bladerunners skating club collected $4,320 from fund-raising activities. They want to buy Ultrablade skates, which are $89 a pair. How many pairs of skates can they buy?

_____

27. An automobile plant produces 1,728 cars in a week. If the plant produces 24 cars each hour, how many hours did it take to produce the cars?

_____

Name _____

Choose the Method: Division

When you divide, you can use pencil and paper, a calculator, or mental math.

Divide. Tell which method you use.

1. 973 ÷ 48 _____

2. 842 ÷ 16 _____

3. 960 ÷ 30 _____

4. 623 ÷ 51 _____

5. 4,984 ÷ 27 _____

6. 3,870 ÷ 90 _____

7. 7,835 ÷ 19 _____

8. 6,940 ÷ 96 _____

9. 27,657 ÷ 75 _____

10. 12,600 ÷ 60 _____

11. 24)89,414 _____

12. 40)80,280 _____

13. 19)612 _____

14. 65)789 _____

15. 82)4,808 _____

16. 60)3,840 _____

17. 39)7,906 _____

18. 84)8,568 _____

19. 36)18,849 _____

20. 73)48,510 _____

21. 30)61,500 _____

22. 36)25,789 _____

Problem Solving
Solve.

23. A train conductor earns $50,596 in a year. How much money does he earn each week?

_____

_____

24. On a trip through Africa, a photographer takes 2,720 pictures. Each roll of film has 32 pictures. How many rolls of film does the photographer use? Show your work.

_____

_____

Use with Grade 5, Chapter 6, Lesson 3, pages 130–131.

Name _____

Divide Decimals by Whole Numbers

Divide. Round each quotient to the nearest hundredth if necessary.

1. $3\overline{)2.19}$

2. $6\overline{)3.63}$

3. $5\overline{)12}$

4. $8\overline{)18.2}$

5. $6\overline{)22}$

6. $4\overline{)2.06}$

7. $8\overline{)16.8}$

8. $10\overline{)118}$

9. $6\overline{)14.23}$

10. $23\overline{)32.2}$

11. $62\overline{)651}$

12. $56\overline{)13.5}$

13. $8.01 \div 9 =$ _____

14. $6.48 \div 40 =$ _____

15. $13.64 \div 7 =$ _____

16. $240.5 \div 64 =$ _____

17. $627 \div 100 =$ _____

18. $26 \div 10 =$ _____

19. $30.87 \div 4 =$ _____

20. $44.4 \div 53 =$ _____

Complete the pattern.

21. $11.7 \div 10 =$ _____

$11.7 \div 100 =$ _____

$11.7 \div 1,000 =$ _____

22. $4.2 \div 10 =$ _____

$4.2 \div 100 =$ _____

$4.2 \div 1,000 =$ _____

23. $89 \div 10 =$ _____

$89 \div 100 =$ _____

$89 \div 1,000 =$ _____

Problem Solving
Solve.

24. Twelve students each ordered a different meal from a fast-food restaurant as part of a science project. When they finished eating, they weighed all the packaging. They found that the packaging weighed a total of 2.88 lb. What was the average weight of the packaging from each meal?

_____

25. Later in the year, the students repeated the experiment exactly. The total weight of the packaging this time was 2.06 lb. To the nearest hundredth of a pound, what was the new average weight of the packaging?

_____

Name _____

Problem Solving: Strategy
Work Backward

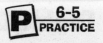

Work backward to solve.

1. Ms. Houston's fifth grade class is going to a dinosaur park. The class raises $68 for the trip. Transportation to the park costs $40. The park sells small fossils for $4 each. How many fossils can they buy with the money they have left?

   _____

2. The outdoors club went on a cross-country ski trip. Rentals for each person cost $4.50. Transportation for the group was $35. The total cost for rentals and transportation was $134. How many rentals did they pay for?

   _____

3. Theresa had $15.65 left after a day at the mall. She spent $35 on a pair of running shoes, $12.50 on a shirt, and $3.85 on lunch. How much money did Theresa have when she arrived at the mall?

   _____

4. **Time** Kusuo's baseball game begins at 5:00 P.M. Kusuo wants to arrive 45 minutes early to warm up. If it takes him $\frac{1}{2}$ hour to get to the baseball field, what time should Kusuo leave his home for the game?

   _____

Mixed Strategy Review
Solve. Use any strategy.

5. A theater seats 44 people. For Friday evening performances, 128 tickets were sold. How many performances were there on Friday evening?

   _____

   Strategy: _____

6. **Science** Many huskies have one brown eye and one blue eye and others have two blue eyes. In a group of 22 huskies, there were 38 blue eyes. How many of the dogs have two blue eyes?

   _____

   Strategy: _____

7. **Number Sense** Steffy picks a number, subtracts 13, and then multiplies the difference by 2. Finally, she adds 8 to the product. Her final number is 122. What was her starting number?

   _____

   Strategy: _____

8. **Create a problem** for which you could work backward to solve. Share it with others.

   _____

   _____

   _____

Use with Grade 5, Chapter 6, Lesson 5, pages 136–137.

Name _____

Explore Collecting, Organizing, and Displaying Data

The Johnson family kept a record of the length
of telephone calls they made in one weekend.

8 minutes	6 minutes	4 minutes	10 minutes	4 minutes	8 minutes
7 minutes	8 minutes	8 minutes	7 minutes	9 minutes	8 minutes
3 minutes	9 minutes	7 minutes	8 minutes	4 minutes	6 minutes
9 minutes	8 minutes	7 minutes	9 minutes	7 minutes	

1. Record the results in the frequency table below.

Length of Calls in Minutes	Number of Calls	
	Tally	Frequency
3		
4		
5		
6		
7		
8		
9		
10		

2. Make a line plot from the frequency table.

Number of Phone Calls

Length of Calls in Minutes

Use data from the line plot for problems 3 and 4.

3. Where does most of the data cluster? What does this tell you?

_____

_____

4. Where is the gap in the line plot? What does this tell you?

_____

_____

Name _____

Range, Mode, Median, and Mean

Find the range, mode, median, and mean.

1. 1, 2, 0, 5, 8, 2, 9, 2, 7 _____

2. 9, 4, 7, 9, 3, 10, 8, 6 _____

3. 34, 17, 10, 23, 21, 15 _____

4. 67, 67, 98, 49, 98, 89 _____

5. 27, 31, 76, 59, 33, 48, 24, 58 _____

6. 105, 126, 90, 50, 75, 90, 62, 112 _____

7. $1.50, $2.50, $1.50, $4.00, $5.00 _____

_____

8. 1.2, 1.5, 2.1, 1.7, 3.2, 2.4, 2.8, 1.3 _____

_____

9. 20, 12.5, 30, 15.4, 25, 18.6, 17.8 _____

_____

10. $3.35, $8.50, $3.35, $4.35, $8.25 _____

_____

11. Find the range, mode, median, and mean.

Student	Ann	Ben	Cara	Fran	Ian	Mike	Kim	Lou
Number of Pets	4	6	0	3	2	5	2	3

_____

Problem Solving
Solve.

During the first five basketball games of the season, a player scored 22, 17, 21, 9, and 17 points.

12. What is the mean of the total points scored per game?

13. How many points must the player score in the next game to make the range 15 points?

_____   _____

Read and Make Pictographs

The table shows the number of books some members of the Kids' Reading Club read one summer.

Number of Books Read During Summer	
Reader	**Number of Books**
Kendra	12
Joel	16
Dan	10
Mae	8
Emily	14

1. Make a pictograph from this data.

Key: _____

_____

2. How did you decide what each picture would represent?

_____

_____

_____

_____

_____

Use data from the pictograph for problems 3–5.

3. How many more minutes a day does Phil read than Mary?

_____

4. What is the total amount of time the club members spend reading in one day?

_____

5. How much time does Zach read each week?

_____

Average Time Spent Reading Each Day	
Greg	📖 📖 📖
Phil	📖 📖 📖 📖 📖
Jessie	📖 📖 📖 📖
Zach	📖 📖
Mary	📖 📖 📖 📖

Key: 📖 = 10 minutes

Name _____

Read and Make Bar Graphs

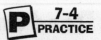

1. The table shows the times Ken and Pat rode their bikes each day last week. Make a double-bar graph of the data.

Time Spent Riding a Bike (in minutes)

Day	Ken	Pat
Sunday	20	25
Monday	30	40
Tuesday	25	20
Wednesday	5	45
Thursday	20	35
Friday	15	35
Saturday	30	20

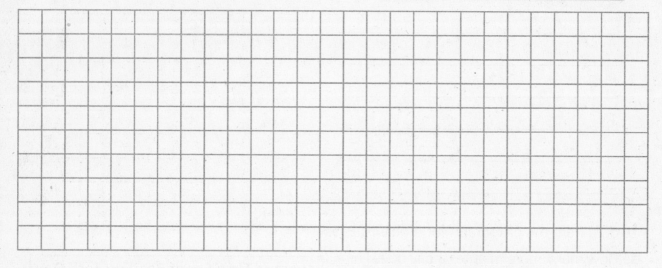

Use data from the graph at right for problems 2–4. **Baseball Playoff Scores**

2. What is the mode of the data?

_____

3. Who won Game 4? By how many runs?

_____

4. The team that wins 3 games wins the playoffs. Who won the playoffs?

_____

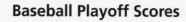

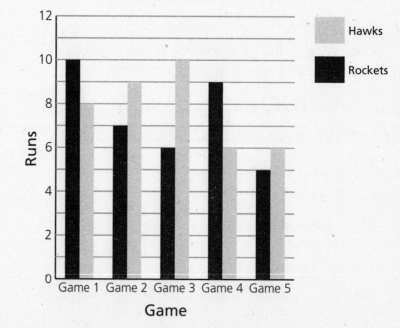

38

Name _____

Read and Make Line Graphs

Use the coordinate grid to name each ordered pair.

1. A _____   **2.** B _____

3. C _____   **4.** D _____

5. E _____   **6.** F _____

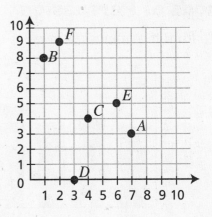

7. Make a line graph to display the data in the table.

Beth's Airplane Trips

Years	Number of Trips
1998	2
1999	3
2000	7
2001	7
2002	5
2003	6

Use the graph at the right for problems 8–10.

8. During how many years did the Martins travel more than 7 days?

_____

9. In which years did the number of travel days increase?

_____

10. In which years did the number of travel days decrease?

_____

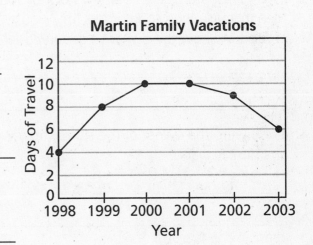

Problem Solving: Skill
Methods of Persuasion

Use data from the graphs to solve.

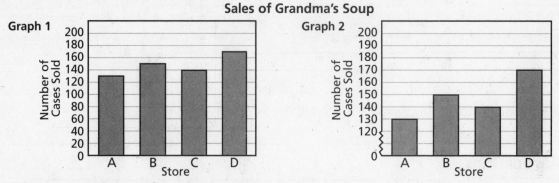

Sales of Grandma's Soup

Graph 1 — Number of Cases Sold (0 to 200), Stores A, B, C, D, Store

Graph 2 — Number of Cases Sold (120 to 200), Stores A, B, C, D, Store

1. A grocery business owns Stores A, B, C, and D. The manager of Store D asks for 40 extra cases of Grandma's Soup with next month's delivery. Which graph do you think he will show to the owner to convince her to increase the order for his store? Explain.

_____

_____

2. Explain why the two graphs look different although they both show the same data.

_____

_____

3. In July each store sells its entire stock of soup. The eager store managers then increase their soup orders for August. Which graph will the manager of Store A show the owner to get an equal share of soup? Explain.

_____

_____

4. If you were the owner of the stores, how would you distribute 100 cases of soup? Explain.

_____

_____

Use with Grade 5, Chapter 7, Lesson 6, pages 168–169.

Name _____

Read and Make Histograms

1. This table shows the distances people rode their bikes on the Riverside Bike Trail one day. Make a histogram to display the data.

Distances Ridden on Trail

Distance in Miles	Number of Cyclists
0–4	9
5–8	16
9–12	24
13–16	14
17–20	6

Use the data from the histogram at the right for problems 2–4.

2. During which time period was the trail the most crowded? The least crowded?

_____

3. Can you tell how many cyclists were on the trail at 5:00 P.M.? Explain.

_____

_____

4. On another day, 3 more cyclists were on the trail at 8 A.M., 5 fewer were on at 2:00 P.M., and 10 more were on at 5:30 P.M. How would the histogram for this day be different?

_____

_____

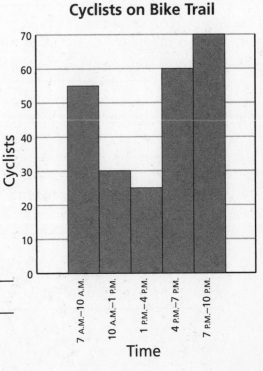

Cyclists on Bike Trail

Use with Grade 5, Chapter 8, Lesson 1, pages 174–176.

Read and Make Stem-and-Leaf Plots

1. The table shows the amount of milk some fifth graders drank in one day. Show the data in a stem-and-leaf plot.

Student	Ounces of Milk Drunk	Student	Ounces of Milk Drunk
Sam	8	Wendi	36
Keiko	32	Van	28
Tonya	24	Sue	32
Sean	12	Marc	21
Phil	20	Anna	32

Milk Drunk in One Day

Key: _____

Use data from the stem-and-leaf plot at the right for problems 2–7.

2. How many students' heights were measured?

_____

3. How many students are taller than 56 inches?

_____

4. What is the height of the tallest student?

_____

5. What is the mode of the students' heights?

_____

6. What is the range of the students' heights?

_____

7. What is the median student height?

_____

Height of Students in Class 101

```
4 | 7 8 8 9 9
5 | 2 4 4 5 7 7 7 8 9 9
6 | 0 0 1 3
```
Key: _____

Use with Grade 5, Chapter 8, Lesson 2, pages 178–179.

Name _____

Problem Solving: Strategy
Make a Graph

For problems 1–4, choose the graph that best displays the data. Explain your choice.

1. The number of walking shoes, running shoes, tennis shoes, and soccer shoes sold in a shoe store.

    A line graph   **B** bar graph

_____

_____

2. The types of sports balls men and women buy in a sports store over a three-month period.

    C histogram   **D** double-bar graph

_____

_____

3. Daily high and low temperatures recorded for a week.

    A double-line graph

    B double-bar graph

_____

_____

4. The favorite composers of the members of a music society.

    C pictograph   **D** histogram

_____

_____

Mixed Strategy Review
Solve. Use any strategy.

5. After buying supplies, Enrico has $11 left. He bought 3 boxes of pens for $5 each and 2 boxes of paper for $18 each. He also bought a cartridge that cost $23. How much money did Enrico have before buying the supplies?

_____

Strategy: _____

6. Art A sculptor has a piece of wood that is 10 feet long. She wants to cut the wood into sections that are 2.5 feet long. How many cuts will she have to make? How many sections will she have?

_____

Strategy: _____

7. Julia and Natalie spent a total of $9.75 for lunch. Julia spent $0.75 more than Natalie. How much did each girl spend for lunch?

_____

Strategy: _____

8. Create a problem in which you must choose the graph that best displays a set of data. Share it with others.

_____

_____

_____

Sampling

Name each population and sample.

1. Survey all the fifth graders in Mr. Smith's class to find how many
students in your school watch a certain TV show.

Population:_____ Sample:_____

2. Survey every third fifth grader who enters the school to
find the most popular TV actor among fifth graders.

Population: _____ Sample: _____

3. Survey customers in an appliance store to find how many
TVs most people in town own.

Population: _____ Sample: _____

4. Survey the customers in Mr. Myers's grocery store for one day
to find how many hours a day most people in town watch TV.

Population: _____ Sample: _____

A researcher wants to find which TV station's news broadcast is
watched by most people in your town. Tell whether or not each sample
is a random sample and whether or not each sample is representative.

5. A survey of all people who visit the zoo on one day

_____

6. A survey of every tenth household from the telephone book

_____

7. A survey of every fifth person who buys a newspaper at a newspaper stand

_____

Problem Solving
Solve.

Brandon wants to find out how many students in his school
must finish their homework before watching TV.

8. He surveys all the students in the cafeteria
during his lunch period. Is the sample a
random sample? Explain.

_____

_____

9. Suggest whom Brandon could survey to
have a random, representative survey.

_____

_____

_____

44

Divisibility

Of 2, 3, 4, 5, 6, 9, and 10, list which numbers each number is divisible by.

1. 87 _____

2. 96 _____

3. 140 _____

4. 423 _____

5. 824 _____

6. 517 _____

7. 210 _____

8. 675 _____

9. 1,293 _____

10. 8,340 _____

11. 4,095 _____

12. 50,006 _____

13. 20,304 _____

14. 86,420 _____

15. 135,952 _____

16. 300,480 _____

17. 8,550 _____

18. 891,235 _____

19. 20 _____

20. 1,592 _____

21. 69,360 _____

22. 9,999 _____

23. 36,521 _____

24. 89,745 _____

25. 2 _____

26. 897,421 _____

Problem Solving
Solve.

27. The school band has 130 members. They march in rows of 5, 6, or 9 members each. Which number of students in each row would make the rows equal?

28. The school chorus has 80 members. They can stand on the stage in rows of 6, 9, or 10. How should the conductor arrange them to have the same number of members in each row?

_____

_____

Name _____

Explore Primes and Composites

Use a factor tree to find the prime factors of each number.

1.   48   **2.**   56   **3.**   36

Write a prime factorization for each number. Use exponents.
Tell if each number is prime or composite.

4. 64   **5.** 45   **6.** 18

_____   _____   _____

_____   _____   _____

7. 23   **8.** 39   **9.** 55

_____   _____   _____

_____   _____   _____

10. 28   **11.** 79   **12.** 62

_____   _____   _____

_____   _____   _____

13. 97   **14.** 88   **15.** 49

_____   _____   _____

_____   _____   _____

Problem Solving
Solve.

16. There are 24 students in Mrs. Green's class. The number of boys and the number of girls are both prime numbers. There are 2 more boys than girls. How many boys and how many girls are in the class?

17. There are 27 students in Mr. Lowell's class. The number of boys and the number of girls are both composite numbers. There are 3 more girls than boys. How many girls and how many boys are in the class?

_____   _____

Common Factors and Greatest Common Factor

Find the GCF of the numbers.

1. 10 and 15 _____

2. 6 and 24 _____

3. 16 and 36 _____

4. 24 and 30 _____

5. 9 and 21 _____

6. 12 and 40 _____

7. 8 and 28 _____

8. 18 and 27 _____

9. 12 and 60 _____

10. 14 and 18 _____

11. 20 and 30 _____

12. 24 and 45 _____

13. 27 and 30 _____

14. 10 and 22 _____

15. 12 and 36 _____

16. 11 and 15 _____

17. 18 and 45 _____

18. 21 and 27 _____

19. 13 and 25 _____

20. 8 and 48 _____

21. 16 and 18 _____

22. 24 and 36 _____

23. 4, 12, and 30 _____

24. 12, 18, and 36 _____

25. 9, 16, and 25 _____

26. 9, 15, and 21 _____

27. 12, 15, and 21 _____

28. 9, 36, and 45 _____

29. 3, 9, and 31 _____

30. 15, 30, and 50 _____

31. 16, 24, and 30 _____

32. 30, 50, and 100 _____

Problem Solving

Solve.

33. Thirty people at the nature center signed up for hiking, and 18 signed up for bird watching. They will be divided up into smaller groups. What is the greatest number of people that can be in each group and have all groups the same size?

_____

34. Rosa found 8 different wildflowers and 20 different leaves on her hike. She plans to display them in 7 equal rows on a poster. What is the greatest number of flowers or leaves she can put in each row?

_____

Name_____

Fractions

Name each fraction shown.

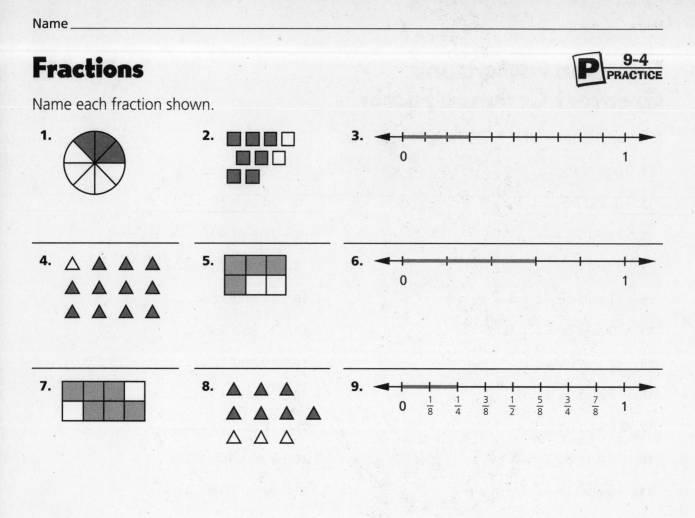

1.

_____

2.

_____

3. 0 _____ 1

_____

4.

_____

5.

_____

6. 0 _____ 1

_____

7.

_____

8.

_____

9. 0 $\frac{1}{8}$ $\frac{1}{4}$ $\frac{3}{8}$ $\frac{1}{2}$ $\frac{5}{8}$ $\frac{3}{4}$ $\frac{7}{8}$ 1

_____

Draw a model to show each fraction.

10. $\frac{3}{5}$

11. $\frac{7}{8}$

12. $\frac{2}{10}$

13. $\frac{4}{6}$

14. $\frac{1}{3}$

15. $\frac{8}{12}$

Problem Solving

Solve.

16. Van has 12 compact discs in his collection. Of these, 7 are by solo performers. What fraction of Van's compact discs are by solo performers?

_____

17. Noah has 9 pennies. Of these, 5 were minted before 2000. What fraction of Noah's pennies were not minted before 2000?

_____

Use with Grade 5, Chapter 9, Lesson 4, pages 206–208.

Name _____

Equivalent Fractions

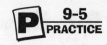

Write two equivalent fractions for each fraction.

1. $\frac{1}{2}$ _____

2. $\frac{1}{4}$ _____

3. $\frac{2}{5}$ _____

4. $\frac{5}{6}$ _____

5. $\frac{7}{8}$ _____

6. $\frac{2}{3}$ _____

7. $\frac{8}{10}$ _____

8. $\frac{3}{8}$ _____

9. $\frac{4}{12}$ _____

10. $\frac{4}{16}$ _____

Algebra Find each missing number.

11. $\frac{1}{4} = \frac{n}{12}$

$n =$ _____

12. $\frac{3}{5} = \frac{a}{10}$

$a =$ _____

13. $\frac{7}{10} = \frac{x}{20}$

$x =$ _____

14. $\frac{8}{12} = \frac{b}{3}$

$b =$ _____

15. $\frac{10}{12} = \frac{y}{6}$

$y =$ _____

16. $\frac{4}{10} = \frac{c}{5}$

$c =$ _____

17. $\frac{3}{5} = \frac{p}{20}$

$p =$ _____

18. $\frac{2}{3} = \frac{6}{a}$

$a =$ _____

19. $\frac{12}{16} = \frac{n}{4}$

$n =$ _____

20. $\frac{9}{15} = \frac{3}{d}$

$d =$ _____

21. $\frac{3}{4} = \frac{j}{24}$

$j =$ _____

22. $\frac{2}{9} = \frac{y}{18}$

$y =$ _____

Problem Solving
Solve.

23. Nina used 24 tiles to make a design. Six of the tiles were blue. Write two equivalent fractions that name the part of the tiles that were blue.

_____

24. Chris walks $\frac{3}{8}$ mile each day to school. Anna walks $\frac{1}{2}$ mile. Do they walk the same distance to school? Explain.

_____

Name _____

Simplify Fractions

Write each fraction in simplest form.

1. $\frac{4}{28}$ _____

2. $\frac{15}{20}$ _____

3. $\frac{6}{21}$ _____

4. $\frac{30}{35}$ _____

5. $\frac{3}{30}$ _____

6. $\frac{12}{14}$ _____

7. $\frac{9}{24}$ _____

8. $\frac{14}{42}$ _____

9. $\frac{20}{25}$ _____

10. $\frac{14}{21}$ _____

11. $\frac{16}{18}$ _____

12. $\frac{4}{36}$ _____

13. $\frac{8}{14}$ _____

14. $\frac{14}{35}$ _____

15. $\frac{10}{12}$ _____

16. $\frac{24}{40}$ _____

17. $\frac{12}{30}$ _____

18. $\frac{4}{32}$ _____

Write each fraction in simplest form. Write *yes* if the fraction is already in simplest form.

19. $\frac{16}{20}$ _____

20. $\frac{1}{2}$ _____

21. $\frac{3}{12}$ _____

22. $\frac{2}{5}$ _____

23. $\frac{3}{7}$ _____

24. $\frac{28}{32}$ _____

25. $\frac{40}{48}$ _____

26. $\frac{12}{18}$ _____

27. $\frac{5}{8}$ _____

28. $\frac{15}{36}$ _____

29. $\frac{2}{3}$ _____

30. $\frac{3}{24}$ _____

31. $\frac{12}{16}$ _____

32. $\frac{9}{10}$ _____

33. $\frac{4}{15}$ _____

Problem Solving
Solve.

34. Of the 27 students in Jarrod's class, 18 receive an allowance each week. What fraction of the students, in simplest form, receive an allowance?

35. Of the 18 students who receive an allowance, 14 do chores around the house. What fraction of these students, in simplest form, do chores around the house?

_____

_____

Use with Grade 5, Chapter 9, Lesson 6, pages 212–213.

Name _____

Problem Solving: Skill
Extra and Missing Information

The table shows the number of students who have volunteered to help with the production of the school play. The school has a budget of $800 to produce the play. Tickets for the play will cost $4.00 for adults. Students will be admitted for free. There are 500 students who attend the school.

Activity	Number of Students
Directors	3
Actors	18
Lighting	6
Sound	4
Special effects	3
Set design	8
Costume design	6
Makeup	2

Solve. If there is extra information, identify the extra information. If there is not enough information, write *not enough information*. Then tell what information you would need to solve the problem.

1. What fraction of the students who attend the school have volunteered to help with the play?

_____

_____

2. What fraction of the volunteers are involved in set design and costume design?

_____

3. If the students sell 300 tickets to the school play, will they have enough money to cover all of the expenses for the play?

_____

_____

4. What fraction of the student volunteers bought tickets for their parents to attend the play?

_____

_____

Name _____

Least Common Multiple and Least Common Denominator

Find the least common multiple (LCM) of the numbers.

1. 5 and 15 _____

2. 2 and 9 _____

3. 2 and 11 _____

4. 6 and 9 _____

5. 4 and 5 _____

6. 8 and 12 _____

7. 4 and 8 _____

8. 10 and 25 _____

9. 3 and 4 _____

10. 2 and 3 _____

11. 8 and 9 _____

12. 4 and 10 _____

13. 2, 4, and 16 _____

14. 3, 5, and 6 _____

15. 3, 6, and 8 _____

Write equivalent fractions using the LCD.

16. $\frac{7}{10}$ and $\frac{2}{5}$ _____

17. $\frac{5}{12}$ and $\frac{1}{4}$ _____

18. $\frac{2}{3}$ and $\frac{3}{8}$ _____

19. $\frac{3}{5}$ and $\frac{9}{10}$ _____

20. $\frac{1}{6}$ and $\frac{7}{12}$ _____

21. $\frac{1}{5}$ and $\frac{2}{3}$ _____

22. $\frac{5}{8}$ and $\frac{2}{5}$ _____

23. $\frac{1}{3}$ and $\frac{5}{12}$ _____

24. $\frac{3}{4}$ and $\frac{13}{16}$ _____

25. $\frac{3}{10}$ and $\frac{5}{6}$ _____

26. $\frac{11}{20}$ and $\frac{4}{5}$ _____

27. $\frac{2}{9}$ and $\frac{1}{8}$ _____

28. $\frac{3}{8}$ and $\frac{5}{6}$ _____

29. $\frac{5}{6}$ and $\frac{9}{24}$ _____

Problem Solving
Solve.

30. José and Sara are walking around the track at the same time. José walks one lap every 8 minutes. Sara walks a lap every 6 minutes. What is the least amount of time they would both have to walk for them to cross the starting point together?

_____

31. Pamela and David walk on the same track. It takes Pamela 9 minutes and David 6 minutes to walk one lap. If they start walking at the same time, how many laps will each have walked when they cross the starting point together for the first time?

_____

Use with Grade 5, Chapter 10, Lesson 1, pages 220–223.

Compare and Order Fractions

Compare. Write >, <, or =.

1. $\frac{3}{4} \bigcirc \frac{7}{12}$

2. $\frac{2}{5} \bigcirc \frac{3}{4}$

3. $\frac{1}{6} \bigcirc \frac{1}{3}$

4. $\frac{1}{2} \bigcirc \frac{7}{10}$

5. $\frac{15}{16} \bigcirc \frac{3}{8}$

6. $\frac{3}{8} \bigcirc \frac{5}{6}$

7. $\frac{7}{8} \bigcirc \frac{8}{9}$

8. $\frac{2}{10} \bigcirc \frac{1}{5}$

9. $\frac{11}{12} \bigcirc \frac{5}{8}$

10. $\frac{4}{5} \bigcirc \frac{17}{20}$

11. $\frac{1}{8} \bigcirc \frac{2}{5}$

12. $\frac{2}{3} \bigcirc \frac{4}{6}$

13. $\frac{1}{5} \bigcirc \frac{1}{4}$

14. $\frac{5}{8} \bigcirc \frac{3}{5}$

15. $\frac{1}{6} \bigcirc \frac{4}{18}$

Order from least to greatest.

16. $\frac{2}{5}, \frac{1}{10}, \frac{3}{20}$ _____

17. $\frac{1}{3}, \frac{1}{9}, \frac{1}{12}$ _____

18. $\frac{3}{8}, \frac{3}{4}, \frac{1}{12}$ _____

19. $\frac{2}{5}, \frac{7}{8}, \frac{4}{5}$ _____

20. $\frac{5}{9}, \frac{5}{8}, \frac{5}{6}$ _____

21. $\frac{5}{8}, \frac{7}{10}, \frac{2}{5}$ _____

22. $\frac{2}{5}, \frac{3}{10}, \frac{1}{4}$ _____

23. $\frac{1}{5}, \frac{2}{15}, \frac{4}{9}$ _____

24. $\frac{7}{12}, \frac{5}{8}, \frac{1}{10}$ _____

25. $\frac{3}{4}, \frac{1}{8}, \frac{5}{16}$ _____

26. $\frac{2}{9}, \frac{2}{3}, \frac{1}{2}$ _____

27. $\frac{3}{5}, \frac{3}{15}, \frac{3}{10}$ _____

Problem Solving

Solve.

28. Each member of the audience in the theater was asked to name a favorite type of play. Drama was named by $\frac{1}{4}$ of the audience, comedy was named by $\frac{11}{20}$, and musical was named by $\frac{1}{5}$. What was the audience's least favorite type of play?

_____

29. Visitors to an art museum were asked to name a favorite type of art. Pottery was named by $\frac{9}{40}$ of the visitors, painting was named by $\frac{2}{5}$, and sculpture was named by $\frac{3}{8}$. What was the favorite type of art of most visitors?

_____

Name _____

Relate Fractions and Decimals

Write each decimal as a fraction in simplest form.

1. 0.3 _____ **2.** 0.49 _____ **3.** 0.7 _____ **4.** 0.50 _____

5. 0.94 _____ **6.** 0.80 _____ **7.** 0.72 _____ **8.** 0.2 _____

9. 0.55 _____ **10.** 0.1 _____ **11.** 0.25 _____ **12.** 0.03 _____

13. 0.77 _____ **14.** 0.6 _____ **15.** 0.26 _____ **16.** 0.99 _____

17. 0.36 _____ **18.** 0.75 _____ **19.** 0.70 _____ **20.** 0.4 _____

21. 0.05 _____ **22.** 0.35 _____ **23.** 0.8 _____ **24.** 0.63 _____

Write each fraction as a decimal.

25. $\frac{2}{5}$ _____ **26.** $\frac{9}{20}$ _____ **27.** $\frac{3}{10}$ _____ **28.** $\frac{7}{25}$ _____

29. $\frac{21}{50}$ _____ **30.** $\frac{1}{2}$ _____ **31.** $\frac{89}{100}$ _____ **32.** $\frac{1}{8}$ _____

33. $\frac{4}{25}$ _____ **34.** $\frac{3}{5}$ _____ **35.** $\frac{23}{25}$ _____ **36.** $\frac{1}{4}$ _____

37. $\frac{17}{20}$ _____ **38.** $\frac{11}{100}$ _____ **39.** $\frac{7}{10}$ _____ **40.** $\frac{3}{8}$ _____

41. $\frac{3}{4}$ _____ **42.** $\frac{5}{8}$ _____ **43.** $\frac{1}{5}$ _____ **44.** $\frac{3}{50}$ _____

45. $\frac{9}{10}$ _____ **46.** $\frac{4}{5}$ _____ **47.** $\frac{1}{20}$ _____ **48.** $\frac{7}{8}$ _____

Problem Solving
Solve.

49. The largest butterfly in the world is found in Papua, New Guinea. The female of the species weighs about 0.9 ounce. Use a fraction to write the female's weight.

_____

50. The shortest recorded fish is the dwarf goby found in the Indo-Pacific. The female of this species is about $\frac{7}{20}$ inch long. Use a decimal to write the female's length.

_____

Use with Grade 5, Chapter 10, Lesson 3, pages 226–227.

Problem Solving: Strategy
Make a Table

Use the Make a Table strategy to solve.

A card shop recorded how many packs of trading cards it sold each week.

Trading Cards Sold					
Week	Number of Packs	Week	Number of Packs	Week	Number of Packs
1	28	5	48	9	25
2	32	6	43	10	37
3	38	7	45	11	42
4	44	8	41	12	35

1. During what fraction of the weeks did the number of packs sold range from 30 to 39? Write the fraction in simplest form.

   _____

2. In what fraction of the weeks were 40 or more packs sold? Write the fraction in simplest form.

   _____

3. **Literature** A bookstore records 8 months of sales of *The Lion, the Witch, and the Wardrobe*, by C.S. Lewis. In what fraction of the months did the number of copies sold range from 20 to 29? Write the fraction in simplest form.

   _____

Bookstore Sales			
Month	Copies	Month	Copies
1	26	5	38
2	24	6	19
3	32	7	15
4	18	8	30

Mixed Strategy Review

Solve. Use any strategy.

4. The number of people who became health club members in May was half as many as the number who became members in June. In July, there were 18 more members than in June. If 76 people became members in July, how many people became members altogether during the three months?

   _____

   Strategy: _____

5. There are 86 students in the school band and the school orchestra. There are 62 students in the band and 46 students in the orchestra. Some students play in both the band and the orchestra. How many students play in both the band and the orchestra?

   _____

   Strategy: _____

Mixed Numbers

Write each mixed number as an improper fraction.

1. $2\frac{3}{4}$ ____

2. $5\frac{1}{6}$ ____

3. $8\frac{1}{2}$ ____

4. $3\frac{2}{3}$ ____

5. $7\frac{2}{5}$ ____

6. $1\frac{9}{10}$ ____

7. $4\frac{7}{8}$ ____

8. $6\frac{5}{7}$ ____

9. $1\frac{8}{9}$ ____

10. $3\frac{12}{17}$ ____

11. $2\frac{1}{10}$ ____

12. $5\frac{5}{13}$ ____

Write each improper fraction as a mixed number in simplest form.

13. $\frac{18}{12}$ ____

14. $\frac{22}{3}$ ____

15. $\frac{27}{9}$ ____

16. $\frac{14}{4}$ ____

17. $\frac{28}{6}$ ____

18. $\frac{64}{8}$ ____

19. $\frac{13}{5}$ ____

20. $\frac{46}{8}$ ____

21. $\frac{21}{8}$ ____

22. $\frac{64}{35}$ ____

23. $\frac{19}{3}$ ____

24. $\frac{44}{8}$ ____

Write each mixed number as a decimal.

25. $3\frac{7}{10}$ ____

26. $4\frac{1}{2}$ ____

27. $4\frac{1}{10}$ ____

28. $5\frac{2}{5}$ ____

29. $8\frac{3}{4}$ ____

30. $2\frac{3}{5}$ ____

31. $5\frac{1}{4}$ ____

32. $1\frac{9}{10}$ ____

Write each decimal as a mixed number in simplest form.

33. 7.5 ____

34. 6.4 ____

35. 5.25 ____

36. 8.31 ____

37. 3.72 ____

38. 2.75 ____

39. 9.6 ____

40. 1.9 ____

41. 5.25 ____

42. 3.9 ____

43. 9.12 ____

44. 6.5 ____

Problem Solving
Solve.

45. A shipment of boxes weighs 30 pounds. There are 8 boxes and each weighs the same number of pounds. How much does each box weigh?

_____

46. Each box in another shipment weighs $3\frac{1}{6}$ pounds. There are 6 boxes in the shipment. What is the total weight of the shipment?

_____

Use with Grade 5, Chapter 10, Lesson 5, pages 230–233.

Compare and Order Fractions, Mixed Numbers, and Decimals

Compare. Write >, <, or =.

1. $3\frac{2}{3}$ ◯ $3\frac{3}{4}$ **2.** 0.75 ◯ $\frac{3}{4}$ **3.** $2\frac{1}{4}$ ◯ 2.4

4. $5\frac{3}{4}$ ◯ 5.825 **5.** 4.3 ◯ $4\frac{3}{5}$ **6.** $4\frac{1}{7}$ ◯ $2\frac{1}{4}$

7. $5\frac{1}{3}$ ◯ 6.7 **8.** 4.25 ◯ $4\frac{1}{3}$ **9.** 0.4 ◯ $\frac{1}{4}$

10. $1\frac{3}{5}$ ◯ $5\frac{7}{10}$ **11.** $3\frac{3}{5}$ ◯ 3.6 **12.** $2\frac{9}{10}$ ◯ 2.99

13. $\frac{3}{4}$ ◯ 0.9 **14.** $7\frac{4}{5}$ ◯ $7\frac{3}{4}$ **15.** $\frac{2}{5}$ ◯ 0.25

16. 9.5 ◯ $9\frac{1}{2}$ **17.** 0.7 ◯ $\frac{4}{5}$ **18.** $8\frac{11}{12}$ ◯ $9\frac{5}{6}$

Order from greatest to least.

19. $\frac{1}{3}, \frac{3}{5}, 0.5$ _____   **20.** $0.2, 0.9, \frac{4}{5}$ _____

21. $\frac{1}{2}, 0.4, \frac{3}{4}$ _____   **22.** $0.6, 5.6, 5\frac{1}{6}$ _____

23. $4.2, 4, 4\frac{1}{4}$ _____   **24.** $2.8, 2\frac{5}{8}, 2\frac{9}{10}$ _____

25. $3.85, 4.65, 3\frac{3}{4}$ _____   **26.** $7\frac{1}{8}, 7.08, 7.18$ _____

27. $4\frac{11}{20}, 3.5, 4\frac{1}{2}$ _____   **28.** $5\frac{1}{4}, 5\frac{1}{3}, 5.28$ _____

Problem Solving
Solve.

29. Martin entered a cross-country race that covered $6\frac{2}{5}$ miles. If Martin runs 6.2 miles, does he complete the race? Explain.

_____

_____

30. Rita tried to break the school high-jump record of $5\frac{3}{4}$ feet. She cleared 5.5 feet. Did she break the record? Explain.

_____

_____

Add Fractions and Mixed Numbers with Like Denominators

Add. Write your answer in simplest form.

1. $\frac{7}{10} + \frac{1}{10} =$ _____

2. $\frac{13}{16} + \frac{7}{16} =$ _____

3. $\frac{4}{5} + \frac{1}{5} =$ _____

4. $\frac{7}{12} + \frac{5}{12} =$ _____

5. $\frac{4}{5} + \frac{3}{5} =$ _____

6. $\frac{5}{6} + \frac{5}{6} =$ _____

7. $\frac{3}{8} + \frac{5}{8} =$ _____

8. $\frac{9}{10} + \frac{7}{10} =$ _____

9. $\frac{3}{4} + \frac{3}{4} =$ _____

10. $3\frac{3}{8}$
$+ 2\frac{1}{8}$
_____

11. $7\frac{2}{3}$
$+ 6\frac{1}{3}$
_____

12. $9\frac{5}{6}$
$+ 5\frac{1}{6}$
_____

13. $11\frac{7}{16}$
$+ 3\frac{3}{16}$
_____

14. $8\frac{3}{10}$
$+ 5\frac{9}{10}$
_____

15. $16\frac{7}{8}$
$+ 7\frac{7}{8}$
_____

16. $4\frac{7}{12}$
$+ 7\frac{11}{12}$
_____

17. $14\frac{19}{20}$
$+ 8\frac{5}{20}$
_____

18. $27\frac{11}{16}$
$+ 43\frac{9}{16}$
_____

19. $98\frac{9}{10}$
$+ 16\frac{1}{10}$
_____

20. $52\frac{1}{6}$
$+ 35\frac{5}{6}$
_____

21. $74\frac{11}{12}$
$+ 29\frac{7}{12}$
_____

Compare. Write >, <, or =.

22. $\frac{7}{8} + \frac{5}{8}$ $\bigcirc$ $\frac{3}{4} + \frac{3}{4}$

23. $\frac{7}{10} + \frac{9}{10}$ $\bigcirc$ $\frac{3}{5} + \frac{4}{5}$

24. $\frac{2}{3} + \frac{2}{3}$ $\bigcirc$ $\frac{5}{12} + \frac{7}{12}$

25. $\frac{3}{8} + \frac{3}{8}$ $\bigcirc$ $\frac{9}{16} + \frac{5}{16}$

26. $2\frac{3}{5} + 1\frac{3}{5}$ $\bigcirc$ $2\frac{7}{10} + 1\frac{7}{10}$

27. $\frac{5}{8} + \frac{7}{8}$ $\bigcirc$ $\frac{13}{16} + \frac{11}{16}$

Problem Solving
Solve.

28. A stock trading at $\$9\frac{5}{8}$ rises $\$1\frac{5}{8}$ during one day. At what price does the stock close that day?

_____

29. Karen knits a scarf. She knit $18\frac{3}{4}$ inches last week. She knit $2\frac{3}{4}$ inches this week. What length is her scarf?

_____

Use with Grade 5, Chapter 11, Lesson 1, pages 252–254.

Problem Solving: Skill
Choose an Operation

Solve. Tell how you chose the operation.

1. Ms. Montoya makes $2\frac{3}{4}$ pounds of goat cheese in the morning. In the afternoon, she makes $1\frac{1}{4}$ pounds of goat cheese. How much goat cheese does Ms. Montoya make for the day?

_____

_____

2. The Wilsons decide to churn butter for a family project. The boys in the family make 2.5 pounds of butter. How much butter do the girls make if the Wilson children make a total of 4.5 pounds of butter?

_____

_____

3. Clara picks 5.75 bushels of apples. Franz picks 3.25 bushels of apples. How many more bushels of apples does Clara pick than Franz?

_____

_____

4. On Monday, Tina makes 4.7 pounds of raisins from grapes. On Tuesday, she makes 3.8 pounds of raisins. How many pounds of raisins does she make in all?

_____

_____

5. Miguel picked $3\frac{1}{4}$ pounds of grapes last week. This week, he picks $2\frac{1}{4}$ pounds of grapes. How many pounds of grapes does Miguel pick altogether?

_____

_____

6. At the beginning of the week, there were 2.85 pounds of almonds in the jar. By the end of the week, there were 1.6 pounds of almonds in the jar. By how many pounds did the almonds decrease during the week?

_____

_____

Explore Adding Fractions with Unlike Denominators

Write the addition sentence shown by each model.
Write the sum in simplest form.

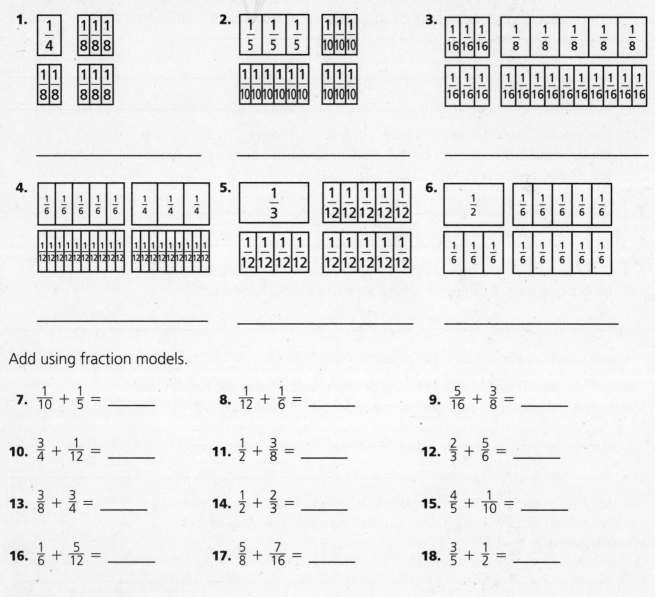

1. _____

2. _____

3. _____

4. _____

5. _____

6. _____

Add using fraction models.

7. $\frac{1}{10} + \frac{1}{5} =$ _____

8. $\frac{1}{12} + \frac{1}{6} =$ _____

9. $\frac{5}{16} + \frac{3}{8} =$ _____

10. $\frac{3}{4} + \frac{1}{12} =$ _____

11. $\frac{1}{2} + \frac{3}{8} =$ _____

12. $\frac{2}{3} + \frac{5}{6} =$ _____

13. $\frac{3}{8} + \frac{3}{4} =$ _____

14. $\frac{1}{2} + \frac{2}{3} =$ _____

15. $\frac{4}{5} + \frac{1}{10} =$ _____

16. $\frac{1}{6} + \frac{5}{12} =$ _____

17. $\frac{5}{8} + \frac{7}{16} =$ _____

18. $\frac{3}{5} + \frac{1}{2} =$ _____

Problem Solving
Solve. Use fraction models.

19. At a park, a picnic shelter covers $\frac{1}{4}$ acre and a playground covers $\frac{5}{8}$ acre. How much area is covered by the picnic shelter and the playground?

_____

20. The tropical rain forest at a zoo covers $\frac{3}{4}$ acre, and the desert area covers $\frac{1}{2}$ acre. How much of the zoo is rain forest and desert?

_____

Name _____

Add Fractions with Unlike Denominators

P 11-4 **PRACTICE**

Add. Write your answer in simplest form.

1. $\dfrac{1}{2}$
$+\ \dfrac{1}{5}$
_____

2. $\dfrac{2}{5}$
$+\ \dfrac{7}{10}$
_____

3. $\dfrac{5}{8}$
$+\ \dfrac{3}{16}$
_____

4. $\dfrac{3}{5}$
$+\ \dfrac{3}{20}$
_____

5. $\dfrac{9}{10}$
$+\ \dfrac{7}{10}$
_____

6. $\dfrac{7}{12}$
$+\ \dfrac{1}{3}$
_____

7. $\dfrac{9}{10}$
$+\ \dfrac{2}{5}$
_____

8. $\dfrac{3}{16}$
$+\ \dfrac{3}{8}$
_____

9. $\dfrac{3}{4}$
$+\ \dfrac{2}{5}$
_____

10. $\dfrac{7}{12}$
$+\ \dfrac{3}{4}$
_____

11. $\dfrac{2}{3}$
$+\ \dfrac{3}{8}$
_____

12. $\dfrac{9}{20}$
$+\ \dfrac{3}{5}$
_____

13. $\dfrac{7}{16} + \dfrac{3}{8} =$ _____

14. $\dfrac{5}{6} + \dfrac{7}{12} =$ _____

15. $\dfrac{15}{16} + \dfrac{5}{8} =$ _____

16. $\dfrac{17}{20} + \dfrac{3}{4} =$ _____

17. $\dfrac{1}{4} + \dfrac{4}{5} =$ _____

18. $\dfrac{1}{2} + \dfrac{1}{5} =$ _____

19. $\dfrac{5}{8} + \dfrac{2}{5} =$ _____

20. $\dfrac{7}{10} + \dfrac{1}{2} =$ _____

21. $\dfrac{5}{6} + \dfrac{5}{8} =$ _____

22. $\dfrac{5}{8} + \dfrac{3}{10} =$ _____

23. $\dfrac{3}{5} + \dfrac{1}{4} =$ _____

24. $\dfrac{5}{6} + \dfrac{7}{9} =$ _____

25. $\dfrac{9}{10} + \dfrac{7}{20} =$ _____

26. $\dfrac{3}{5} + \dfrac{5}{6} =$ _____

27. $\dfrac{5}{8} + \dfrac{35}{12} =$ _____

Problem Solving
Solve.

28. After school, Michael walks $\dfrac{3}{5}$ mile to the park and then walks $\dfrac{3}{4}$ mile to his house. How far does Michael walk from school to his house?

_____

29. When Rachel walks to school on the sidewalk, she walks $\dfrac{7}{10}$ mile. When she takes the shortcut across the field, she walks $\dfrac{1}{4}$ mile less. How long is the shorter route?

_____

Name_____

Add Mixed Numbers with Unlike Denominators

Add. Write your answer in simplest form.

1. $5\frac{3}{4} + 4\frac{1}{12} =$ _____   **2.** $3\frac{1}{3} + 7\frac{7}{10} =$ _____   **3.** $2\frac{11}{12} + 9\frac{5}{8} =$ _____

4. $7\frac{9}{10} + \frac{3}{5} =$ _____   **5.** $6\frac{1}{6} + 6\frac{3}{4} =$ _____   **6.** $5\frac{2}{5} + 8\frac{3}{8} =$ _____

7. $14\frac{7}{16} + 25\frac{3}{4} =$ _____   **8.** $38\frac{3}{8} + 19\frac{13}{16} =$ _____   **9.** $46 + 37\frac{11}{12} =$ _____

10. $\begin{array}{r} 3\frac{5}{12} \\ + 4\frac{1}{6} \\ \hline \end{array}$ **11.** $\begin{array}{r} 5\frac{1}{2} \\ + 9\frac{3}{8} \\ \hline \end{array}$ **12.** $\begin{array}{r} 7\frac{5}{8} \\ + 2\frac{3}{4} \\ \hline \end{array}$ **13.** $\begin{array}{r} 1\frac{3}{5} \\ + 8\frac{1}{4} \\ \hline \end{array}$

14. $\begin{array}{r} 6\frac{1}{4} \\ + 7\frac{9}{10} \\ \hline \end{array}$ **15.** $\begin{array}{r} \frac{5}{6} \\ + 3\frac{3}{8} \\ \hline \end{array}$ **16.** $\begin{array}{r} 7\frac{2}{3} \\ + 8\frac{3}{5} \\ \hline \end{array}$ **17.** $\begin{array}{r} 5\frac{7}{10} \\ + 6\frac{3}{4} \\ \hline \end{array}$

18. $\begin{array}{r} 24\frac{3}{16} \\ + 32\frac{5}{8} \\ \hline \end{array}$ **19.** $\begin{array}{r} 56\frac{13}{20} \\ + 19\frac{4}{5} \\ \hline \end{array}$ **20.** $\begin{array}{r} 37\frac{2}{3} \\ + 45\frac{5}{8} \\ \hline \end{array}$ **21.** $\begin{array}{r} 18\frac{7}{12} \\ + 76\frac{3}{10} \\ \hline \end{array}$

Problem Solving
Solve.

22. A house is $52\frac{3}{4}$ feet wide. The attached garage is $20\frac{1}{2}$ feet wide. What is the total width of the house?

_____

23. In a family room, a fireplace is $12\frac{5}{6}$ feet wide. The total wall space on the sides of the fireplace is $18\frac{1}{4}$ feet wide. How wide is the family room?

_____

Name_____

Algebra: Properties of Addition

P 11-6 PRACTICE

Find each missing number. Identify the property you used.

1. $\frac{3}{4} + \boxed{} = \frac{3}{4}$

2. $3\frac{1}{8} + (\frac{5}{8} + 1\frac{1}{4}) = (3\frac{1}{8} + \boxed{}) + 1\frac{1}{4}$

_____

_____

3. $\frac{9}{16} + \frac{1}{4} = \frac{1}{4} + \boxed{}$

4. $\frac{1}{10} + (\frac{3}{5} + \boxed{}) = (\frac{1}{10} + \frac{3}{5}) + \frac{1}{2}$

_____

_____

5. $7\frac{1}{2} = \boxed{} + 0$

6. $2\frac{1}{3} + (1\frac{1}{2} + \frac{1}{3}) = 2\frac{1}{3} + (\boxed{} + 1\frac{1}{2})$

_____

_____

Use the Associative Property to solve. Show your work.

7. $\frac{1}{4} + (\frac{1}{4} + \frac{1}{5})$

8. $(\frac{1}{6} + \frac{3}{8}) + \frac{7}{8}$

_____

_____

9. $(\frac{1}{2} + 4\frac{1}{10}) + \frac{7}{10}$

10. $\frac{5}{12} + (2\frac{7}{12} + 1\frac{5}{6})$

_____

_____

11. $(2\frac{5}{8} + 3\frac{1}{2}) + 4\frac{1}{2}$

12. $3\frac{7}{16} + (1\frac{5}{16} + 3\frac{1}{3})$

_____

_____

Algebra Find each missing number.

13. $\frac{3}{8} + \frac{7}{12} = c + \frac{3}{8}$

14. $2\frac{1}{5} = h + 2\frac{1}{5}$

15. $3\frac{7}{10} + 0 = 0 + s$

$c = $ _____

$h = $ _____

$s = $ _____

Problem Solving

Solve.

16. During one day at school, Marc spends $4\frac{3}{4}$ hours in class. He also spends $\frac{2}{3}$ hour at lunch and $\frac{3}{4}$ hour at recess. How long is Marc's school day?

_____

17. One afternoon, Laura spent $\frac{1}{2}$ hour doing math homework, $\frac{3}{4}$ hour doing English homework, and $1\frac{1}{4}$ hours doing science homework. How much time did she spend doing homework that afternoon?

_____

Name _____

Subtract Fractions and Mixed Numbers with Like Denominators

P **PRACTICE** 12-1

Subtract. Write your answer in simplest form.

1. $\frac{13}{16} - \frac{7}{16} =$ _____

2. $\frac{7}{8} - \frac{1}{8} =$ _____

3. $\frac{11}{12} - \frac{7}{12} =$ _____

4. $1\frac{5}{8} - 1\frac{3}{8} =$ _____

5. $4\frac{7}{12} - 1\frac{1}{12} =$ _____

6. $9\frac{5}{8} - 3\frac{1}{8} =$ _____

7. $\begin{array}{r} \frac{7}{15} \\ -\ \frac{2}{15} \\ \hline \end{array}$

8. $\begin{array}{r} \frac{9}{20} \\ -\ \frac{3}{20} \\ \hline \end{array}$

9. $\begin{array}{r} 9\frac{3}{8} \\ -\ 5\frac{1}{8} \\ \hline \end{array}$

10. $\begin{array}{r} 8\frac{11}{12} \\ -\ 2\frac{1}{12} \\ \hline \end{array}$

11. $\begin{array}{r} 25\frac{11}{20} \\ -\ 12\frac{7}{20} \\ \hline \end{array}$

12. $\begin{array}{r} 7\frac{9}{16} \\ -\ 6\frac{7}{16} \\ \hline \end{array}$

13. $\begin{array}{r} 10\frac{4}{5} \\ -\ 7\frac{3}{5} \\ \hline \end{array}$

14. $\begin{array}{r} 52\frac{7}{9} \\ -\ 16\frac{4}{9} \\ \hline \end{array}$

Algebra Find each missing number.

15. $\frac{2}{5} + n = 1\frac{1}{5}$

$n =$ _____

16. $\frac{7}{8} - t = \frac{3}{4}$

$t =$ _____

17. $s - \frac{5}{16} = \frac{1}{4}$

$s =$ _____

18. $r + 2\frac{1}{4} = 4$

$r =$ _____

19. $8\frac{1}{6} - b = 7$

$b =$ _____

20. $3\frac{11}{12} + h = 6\frac{1}{3}$

$h =$ _____

Problem Solving
Solve.

21. Wilma pitches $4\frac{2}{3}$ innings in a baseball game. Nina pitches $1\frac{1}{3}$ innings in the same game. How many more innings does Wilma pitch than Nina?

_____

22. Robert lives $3\frac{3}{10}$ miles from school. Al lives $4\frac{7}{10}$ miles from school. Who lives farther from school? How much farther?

_____

Use with Grade 5, Chapter 12, Lesson 1, pages 272–275.

PRACTICE

© Macmillan/McGraw-Hill. All rights reserved.

Subtract Fractions with Unlike Denominators

Write the subtraction sentence shown by each model. Write the difference in simplest form.

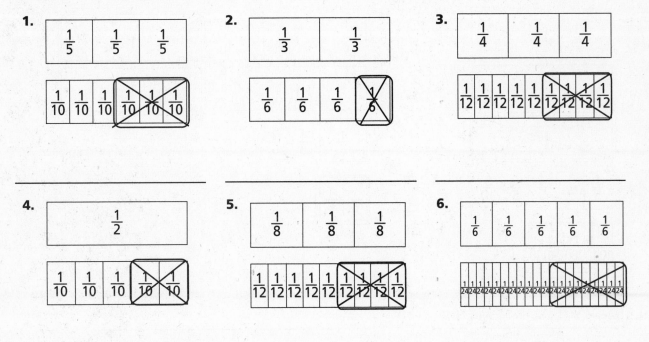

1.

2.

3.

4.

5.

6.

Subtract. Write your answer in simplest form.

7. $\frac{7}{12} - \frac{1}{4} =$ _____

8. $\frac{1}{2} - \frac{1}{3} =$ _____

9. $\frac{9}{10} - \frac{2}{5} =$ _____

10. $\frac{5}{8} - \frac{1}{4} =$ _____

11. $\frac{11}{20} - \frac{3}{10} =$ _____

12. $\frac{11}{12} - \frac{1}{3} =$ _____

13. $\frac{7}{10} - \frac{1}{2} =$ _____

14. $\frac{3}{4} - \frac{2}{3} =$ _____

15. $\frac{5}{6} - \frac{3}{4} =$ _____

16. $\frac{3}{4} - \frac{3}{5} =$ _____

17. $\frac{11}{12} - \frac{1}{4} =$ _____

18. $\frac{4}{5} - \frac{1}{2} =$ _____

Problem Solving

Solve.

19. The distance around a lily pond is $\frac{7}{10}$ mile. Rocks have been placed for $\frac{1}{4}$ mile along the pond's edge. How much of the edge does not have rocks?

_____

20. The first $\frac{1}{5}$ mile of a $\frac{3}{4}$-mile path through a rose garden is paved with bricks. How much of the path is not paved with bricks?

_____

Name _____

Explore Subtracting Mixed Numbers with Unlike Denominators

Write the subtraction sentence shown by each model.
Write the answer in simplest form.

1.

| 1 |
| 1 |
| 1 |

| $\frac{1}{5}$ | $\frac{1}{5}$ | $\frac{1}{5}$ | $\frac{3}{5} > \frac{5}{10}$ |

| $\frac{1}{10}$ | $\frac{1}{10}$ | $\frac{1}{10}$ | $\frac{1}{10}$ | $\frac{1}{10}$ | $\frac{1}{10}$ |

2.

| 1 |
| 1 |

| 1 |
| 1 |

| $\frac{1}{3}$ | $\frac{1}{5} = \frac{2}{6}$ |

| $\frac{1}{6}$ | $\frac{1}{6}$ |

_____   _____

Subtract using models. Write your answer in simplest form.

3. $3\frac{7}{8} - 2\frac{1}{4} =$ _____

4. $4\frac{5}{6} - 1\frac{1}{3} =$ _____

5. $2\frac{9}{10} - 2\frac{3}{5} =$ _____

6. $3\frac{3}{4} - 1\frac{5}{16} =$ _____

7. $2\frac{1}{2} - 1\frac{3}{8} =$ _____

8. $4\frac{11}{12} - 2\frac{5}{6} =$ _____

9. $4\frac{1}{6} - 2\frac{1}{2} =$ _____

10. $3\frac{7}{10} - 2\frac{1}{5} =$ _____

11. $3\frac{3}{8} - 1\frac{3}{4} =$ _____

12. $3\frac{5}{16} - 2\frac{5}{8} =$ _____

13. $4\frac{1}{3} - 1\frac{1}{2} =$ _____

14. $3\frac{5}{12} - 2\frac{2}{3} =$ _____

Problem Solving
Solve.

15. A disc jockey has $4\frac{3}{4}$ minutes of radio time to fill with a song and a commercial. If the song lasts $3\frac{1}{2}$ minutes, how much time remains for the commercial?

16. A listener requests a disc jockey to play a song that lasts $3\frac{3}{4}$ minutes. Only $2\frac{5}{6}$ minutes of time is available. How long before the end of the song will the music have to stop?

_____   _____

Use with Grade 5, Chapter 12, Lesson 3, pages 280–281.

Name_____

Subtract Mixed Numbers

Subtract. Write your answer in simplest form.

1. $7\frac{15}{16} - 2\frac{11}{16} =$ _____

2. $11\frac{4}{5} - 4\frac{3}{10} =$ _____

3. $12 - 9\frac{1}{3} =$ _____

4. $18\frac{1}{6} - 9\frac{5}{6} =$ _____

5. $9 - 5\frac{1}{12} =$ _____

6. $16\frac{1}{3} - 7\frac{7}{10} =$ _____

7. $34\frac{11}{20} - 15 =$ _____

8. $64\frac{3}{4} - 37\frac{11}{12} =$ _____

9. $51\frac{2}{5} - 25\frac{3}{4} =$ _____

10. $46 - 27\frac{3}{4} =$ _____

11. $82\frac{4}{5} - 62 =$ _____

12. $23\frac{1}{8} - 15\frac{2}{5} =$ _____

13. $16 - 7\frac{11}{12} =$ _____

14. $35\frac{7}{8} - 21\frac{1}{4} =$ _____

15. $97 - 87\frac{4}{5} =$ _____

16. $\quad 6\frac{11}{12}$
$\underline{- 4\frac{5}{12}}$

17. $\quad 11\frac{2}{3}$
$\underline{- 3\frac{2}{5}}$

18. $\quad 14\frac{7}{8}$
$\underline{- 5\quad}$

19. $\quad 15\frac{1}{6}$
$\underline{- 6\frac{1}{4}}$

20. $\quad 9\frac{3}{10}$
$\underline{- 8\frac{7}{10}}$

21. $\quad 12\frac{1}{2}$
$\underline{- 3\frac{1}{5}}$

22. $\quad 44$
$\underline{- 21\frac{13}{16}}$

23. $\quad 74\frac{3}{8}$
$\underline{- 38\frac{3}{5}}$

24. $\quad 50\frac{1}{2}$
$\underline{- 41\quad}$

25. $\quad 35\frac{3}{8}$
$\underline{- 18\frac{3}{4}}$

26. $\quad 99\frac{9}{10}$
$\underline{- 75\frac{3}{5}}$

27. $\quad 23$
$\underline{- 14\frac{5}{12}}$

Problem Solving
Solve.

28. A grocery bag will hold 8 pounds of oranges. Kyle puts $3\frac{5}{8}$ pounds of oranges in the bag. How many more pounds of oranges can he put in the bag?

_____

29. Sara needs $2\frac{1}{2}$ pounds of grapes for a salad. She buys a bag of grapes that weighs only $1\frac{7}{8}$ pounds. How many more pounds of grapes does she need?

_____

_____

Problem Solving: Strategy
Write an Equation

PRACTICE 12-5

Write an equation, then solve.

1. At the end of the day, a baker has $3\frac{1}{2}$ pounds of rye flour left. How many pounds of rye flour does he use if he starts with 15 pounds?

   _____

2. A chef buys 50 pounds of rice. At the end of the week, she has $12\frac{3}{4}$ pounds of rice left. How much does she use?

   _____

3. **Social Studies** Daniel Boone was born in 1734. In 1789, George Washington was elected the first president of the United States. How old was Daniel Boone when Washington was elected President?

   _____

4. **Measurement** A carpenter has a piece of wood 12 feet long. After he cuts the wood into pieces, $3\frac{5}{8}$ feet are left. How much of the wood does the carpenter use?

   _____

Mixed Strategy Review
Solve. Use any strategy.

5. Five students are lined up in the cafeteria. Beth is first in line. Jeff is 2 places behind Ernesto. Leah is ahead of Peter, who is fifth in line. Who is third in line?

   _____

   Strategy: _____

6. **Time** Amanda leaves her house at 4:12 P.M. She meets Kyle 6 minutes later and together they walk to the library. If they arrive at the library at 4:35 P.M., how long do Kyle and Amanda walk together?

   _____

   Strategy: _____

7. Daniel makes candles. The supplies needed to make a dozen candles cost $5.25. How much profit does he make if he sells all the candles for $4.50 each?

   _____

   Strategy: _____

8. **Write a problem** for which you could write an equation to solve. Share it with others.

   _____

   _____

   _____

© Macmillan/McGraw-Hill. All rights reserved.

68

Use with Grade 5, Chapter 12, Lesson 5, pages 284–285.

Estimate Sums and Differences of Mixed Numbers

Round to the nearest whole number.

1. $7\frac{3}{4}$ _____

2. $4\frac{1}{6}$ _____

3. $8\frac{3}{10}$ _____

4. $3\frac{1}{2}$ _____

5. $2\frac{9}{16}$ _____

6. $9\frac{4}{5}$ _____

7. $1\frac{7}{8}$ _____

8. $5\frac{5}{12}$ _____

Estimate.

9. $3\frac{7}{8} + 2\frac{1}{6}$

10. $8\frac{5}{6} - 3\frac{2}{3}$

11. $5\frac{1}{8} - 1\frac{7}{8}$

12. $9\frac{7}{10} + 3\frac{4}{5}$

13. $6\frac{1}{4} + 7\frac{3}{8}$

14. $14\frac{1}{5} - 9\frac{3}{5}$

15. $18\frac{5}{16} - 9\frac{13}{16}$

16. $6\frac{11}{12} + 4\frac{5}{12}$

17. $7\frac{1}{3} + 6\frac{7}{12}$

18. $15\frac{3}{8} - 7\frac{7}{16}$

19. $9\frac{4}{5} + 6\frac{2}{3}$

20. $6\frac{11}{12} - 6\frac{1}{5}$

21. $8\frac{2}{5} + 8\frac{11}{16}$

22. $17\frac{7}{10} - 9\frac{1}{3}$

23. $7\frac{1}{2} + 9\frac{3}{8}$

24. $25\frac{7}{12} + 34\frac{1}{12}$

25. $58\frac{4}{5} - 29\frac{7}{8}$

26. $52\frac{1}{3} - 34\frac{5}{16}$

Problem Solving

Solve.

27. Beth walks $10\frac{7}{8}$ miles in one week. She walks $2\frac{1}{2}$ fewer miles the following week. About how many miles does she walk the second week?

28. Jon wants to walk at least 8 miles by the end of the week. He walks $5\frac{3}{4}$ miles by Thursday. If he walks another $2\frac{5}{8}$ miles on Friday, will he meet his goal? Explain.

Multiply a Whole Number by a Fraction

Multiply. Write your answer in simplest form.

1. $\frac{1}{5} \times 45 =$ _____

2. $\frac{5}{8} \times 32 =$ _____

3. $\frac{3}{4} \times 40 =$ _____

4. $\frac{1}{2} \times 14 =$ _____

5. $\frac{4}{9} \times 63 =$ _____

6. $\frac{2}{11} \times 33 =$ _____

7. $\frac{3}{10} \times 70 =$ _____

8. $\frac{5}{7} \times 42 =$ _____

9. $\frac{1}{10} \times 20 =$ _____

10. $\frac{2}{5} \times 35 =$ _____

11. $\frac{9}{10} \times 50 =$ _____

12. $\frac{5}{8} \times 24 =$ _____

13. $\frac{1}{12} \times 96 =$ _____

14. $\frac{8}{9} \times 72 =$ _____

15. $\frac{1}{3} \times 18 =$ _____

16. $\frac{7}{9} \times 45 =$ _____

17. $\frac{3}{5} \times 15 =$ _____

18. $\frac{3}{4} \times 48 =$ _____

19. $\frac{7}{10} \times 30 =$ _____

20. $\frac{4}{7} \times 77 =$ _____

21. $\frac{7}{8} \times 64 =$ _____

22. $\frac{2}{3} \times 27 =$ _____

23. $\frac{5}{12} \times 60 =$ _____

24. $\frac{3}{11} \times 88 =$ _____

Algebra Complete each table.

25. Rule: Multiply by $\frac{1}{4}$

Input	Output
12	
20	
32	
40	

26. Rule: Multiply by $\frac{5}{6}$

Input	Output
12	
30	
42	
54	

27. Rule: Multiply by $\frac{7}{12}$

Input	Output
36	
60	
96	
144	

Problem Solving

Solve.

28. A basketball that normally sells for $24 is on sale for $\frac{2}{3}$ of the regular price. What is the sale price of the basketball?

_____

29. Kari received a $30 gift certificate. After she bought a sweatshirt, she had $\frac{3}{10}$ of the money left. How much money did Kari have left?

_____

Use with Grade 5, Chapter 13, Lesson 1, pages 302–303.

Explore Multiplication of a Fraction by a Fraction

Use the model to find each answer.

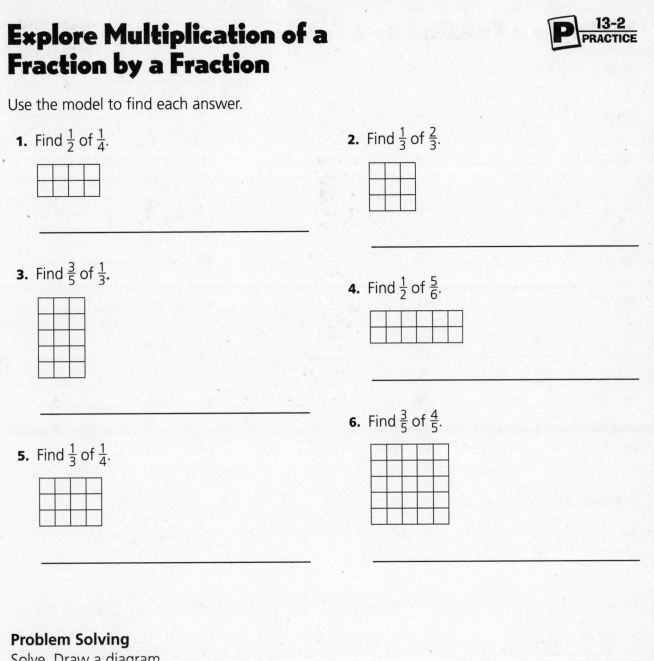

1. Find $\frac{1}{2}$ of $\frac{1}{4}$.

_____

2. Find $\frac{1}{3}$ of $\frac{2}{3}$.

_____

3. Find $\frac{3}{5}$ of $\frac{1}{3}$.

_____

4. Find $\frac{1}{2}$ of $\frac{5}{6}$.

_____

5. Find $\frac{1}{3}$ of $\frac{1}{4}$.

_____

6. Find $\frac{3}{5}$ of $\frac{4}{5}$.

_____

Problem Solving

Solve. Draw a diagram.

7. A farmer plants crops on $\frac{1}{2}$ of his land. Of the land that has crops, $\frac{1}{3}$ of the land has corn. What part of the farm is planted with corn?

_____

Name _____

Multiply a Fraction by a Fraction

Multiply. Write each answer in simplest form.

1. $\frac{1}{2} \times \frac{3}{8} =$ _____

2. $\frac{7}{12} \times \frac{4}{5} =$ _____

3. $\frac{3}{4} \times \frac{1}{9} =$ _____

4. $\frac{4}{9} \times \frac{5}{6} =$ _____

5. $\frac{3}{4} \times \frac{1}{3} =$ _____

6. $\frac{5}{8} \times \frac{3}{10} =$ _____

7. $\frac{2}{9} \times \frac{1}{2} =$ _____

8. $\frac{3}{5} \times \frac{3}{8} =$ _____

9. $\frac{8}{9} \times \frac{5}{16} =$ _____

10. $\frac{1}{5} \times \frac{7}{12} =$ _____

11. $\frac{3}{10} \times \frac{1}{4} =$ _____

12. $\frac{5}{7} \times \frac{7}{9} =$ _____

13. $\frac{9}{20} \times \frac{2}{3} =$ _____

14. $\frac{3}{5} \times \frac{7}{12} =$ _____

15. $\frac{1}{16} \times \frac{8}{9} =$ _____

16. $\frac{2}{3} \times \frac{3}{5} =$ _____

17. $\frac{2}{7} \times \frac{13}{20} =$ _____

18. $\frac{4}{5} \times \frac{7}{16} =$ _____

19. $\frac{11}{12} \times \frac{6}{7} =$ _____

20. $\frac{2}{3} \times \frac{7}{8} =$ _____

21. $\frac{1}{6} \times \frac{1}{12} =$ _____

22. $\frac{5}{9} \times \frac{3}{20} =$ _____

23. $\frac{9}{16} \times \frac{2}{3} =$ _____

24. $\frac{3}{20} \times \frac{4}{5} =$ _____

Algebra   Find *n* so that each expression is true.

25. $\frac{1}{6} \times \frac{n}{2} = \frac{1}{12}$

$n =$ _____

26. $\frac{5}{6} \times \frac{n}{8} = \frac{15}{48}$

$n =$ _____

27. $\frac{7}{8} \times \frac{n}{8} = \frac{35}{64}$

$n =$ _____

28. $\frac{2}{3} \times \frac{n}{8} = \frac{7}{12}$

$n =$ _____

29. $\frac{4}{5} \times \frac{n}{4} = \frac{3}{5}$

$n =$ _____

30. $\frac{3}{4} \times \frac{n}{6} = \frac{5}{8}$

$n =$ _____

Problem Solving
Solve.

31. Each year the Gardners plant $\frac{7}{8}$ of an acre with tomatoes. They sell half of what they grow at a roadside stand. What part of an acre do the Gardners use for the tomatoes they sell?

32. The Wilsons' garden covers $\frac{5}{8}$ acre. One fourth of the garden is planted with flowers. The rest is vegetables. What part of an acre is planted with flowers? With vegetables?

_____

_____

Use with Grade 5, Chapter 13, Lesson 3, pages 306–308.

Name_____

Problem Solving: Skill
Multi-step Problems

Solve. Use the diagrams below to solve problems 1–5.

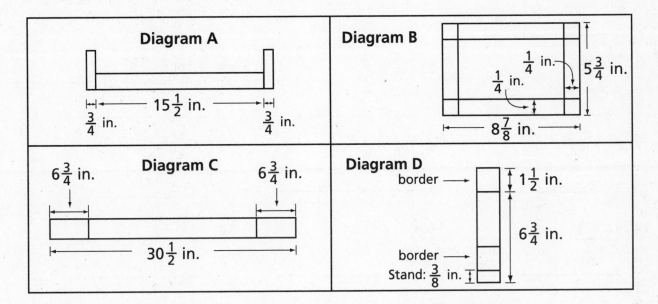

1. Diagram A shows a view of a plan for a bookcase. Books will sit on the interior of the bookcase. What is the length of the bottom of the bookcase?

_____

2. Diagram B shows a plan for a rectangular frame. What is the greatest width a picture can have to fit inside the frame?

_____

3. Diagram B shows a plan for a rectangular frame. What is the greatest length a picture can have to fit inside the frame?

_____

4. Diagram C shows a plan for the design of a folding leaf table. The sections at each end show the leaves that fold down. How long is the table when the leaves are folded down?

_____

5. Diagram D shows the plan for a countertop spice rack that sits on a stand. How tall is the spice rack without the stand?

_____

Estimate Products

Estimate.

1. $\frac{1}{2} \times 13$

2. $7 \times 3\frac{1}{4}$

3. $\frac{4}{7} \times 8\frac{1}{9}$

_____

_____

_____

4. $\frac{5}{6} \times 23$

5. $21\frac{8}{9} \times \frac{5}{12}$

6. $17 \times \frac{2}{5}$

_____

_____

_____

7. $2\frac{1}{6} \times 9\frac{3}{4}$

8. $13\frac{7}{8} \times \frac{3}{8}$

9. $6 \times 8\frac{4}{5}$

_____

_____

_____

10. $31 \times \frac{2}{3}$

11. $\frac{2}{5} \times 24\frac{1}{4}$

12. $3\frac{5}{6} \times 4\frac{2}{3}$

_____

_____

_____

13. $\frac{7}{8} \times 62$

14. $1\frac{11}{12} \times 9\frac{1}{5}$

15. $34 \times \frac{1}{6}$

_____

_____

_____

16. $5\frac{7}{9} \times 4$

17. $\frac{5}{12} \times 49$

18. $23\frac{3}{8} \times 42\frac{7}{9}$

_____

_____

_____

Estimate to compare. Write >, <, or =.

19. $34 \times \frac{3}{4}$ ◯ $59\frac{5}{6} \times \frac{4}{9}$ **20.** $\frac{3}{8} \times 33$ ◯ $\frac{5}{8} \times 10\frac{1}{4}$ **21.** $57\frac{1}{2} \times 18\frac{3}{5}$ ◯ $37\frac{5}{6} \times 27\frac{1}{3}$

_____

_____

_____

Problem Solving

Solve.

22. Teresa rode $6\frac{7}{10}$ miles on her bike in one hour. If she continues at this pace, about how far could she ride in 5 hours?

23. Chan is riding his bike on a 48-mile cross-country course. He knows that $\frac{2}{5}$ of the course is uphill. About how many miles will Chan have to ride uphill?

_____

_____

Use with Grade 5, Chapter 13, Lesson 5, pages 312–314.

Choose the Method: Multiply Fractions

When you multiply with fractions, you can use pencil and paper, a calculator, or mental math.

Multiply and write your answer in simplest form. Tell which method you use.

1. $\frac{5}{6} \times 24 =$ _____

2. $\frac{3}{7} \times \frac{5}{8} =$ _____

3. $80 \times \frac{3}{4} =$ _____

4. $\frac{5}{12} \times \frac{3}{5} =$ _____

5. $\frac{7}{8} \times 96 =$ _____

6. $\frac{3}{5} \times \frac{4}{9} =$ _____

7. $\frac{1}{2} \times \frac{3}{5} \times 40 =$ _____

8. $\frac{7}{10} \times 200 \times \frac{1}{4} =$ _____

9. $\frac{1}{3} \times \frac{3}{8} \times 24 =$ _____

10. $\frac{4}{7} \times 42 \times \frac{1}{8} =$ _____

11. $128 \times \frac{3}{4} \times \frac{1}{2} =$ _____

12. $\frac{5}{6} \times 120 \times \frac{1}{4} =$ _____

13. $\frac{1}{3} \times \frac{3}{11} \times 110 =$ _____

14. $\frac{5}{9} \times \frac{3}{7} \times \frac{1}{5} =$ _____

Algebra Find each missing number.

15. $\frac{n}{9} \times \frac{3}{4} = \frac{5}{12}$ $n =$ _____

16. $\frac{7}{8} \times \frac{1}{a} = \frac{7}{16}$ $a =$ _____

17. $150 \times \frac{v}{5} = 120$ $v =$ _____

18. $\frac{4}{9} \times u = 4$ $u =$ _____

19. $\frac{3}{10} \times \frac{5}{y} = \frac{3}{16}$ $y =$ _____

20. $\frac{p}{3} \times 180 = 120$ $p =$ _____

Problem Solving
Solve.

21. A book has 450 pages. Robin has read $\frac{3}{5}$ of the book. How many pages has Robin read?

_____

22. The stands at the soccer field can hold 400 people. The stands are $\frac{3}{4}$ filled. Of the people in the stands, $\frac{5}{6}$ are home team fans. How many home-team fans are at the game?

_____

Use with Grade 5, Chapter 13, Lesson 6, pages 316–317.

Name _____

Multiply a Mixed Number by a Whole Number

Multiply. Write each answer in simplest form.

1. $2\frac{1}{8} \times 4 =$ _____

2. $7\frac{2}{3} \times 9 =$ _____

3. $4\frac{5}{6} \times 3 =$ _____

4. $3\frac{1}{4} \times 4 =$ _____

5. $8 \times 6\frac{7}{10} =$ _____

6. $3 \times 7\frac{8}{9} =$ _____

7. $24 \times 3\frac{3}{4} =$ _____

8. $40 \times 1\frac{3}{8} =$ _____

9. $4\frac{1}{12} \times 30 =$ _____

10. $9 \times 2\frac{3}{5} =$ _____

11. $6 \times 2\frac{3}{10} \times 5 =$ _____

12. $7\frac{2}{3} \times 20 \times 3 =$ _____

13. $9 \times 5\frac{3}{4} \times 8 =$ _____

14. $16 \times 4 \times 2\frac{9}{16} =$ _____

Algebra Find the number that makes each sentence true.

15. $n \times 4\frac{7}{8} = 14\frac{5}{8}$ $n =$ _____

16. $6\frac{1}{3} \times a = 31\frac{2}{3}$ $a =$ _____

17. $2\frac{4}{5} \times v = 16\frac{4}{5}$ $v =$ _____

18. $u \times 7\frac{9}{10} = 63\frac{1}{5}$ $u =$ _____

19. $11\frac{1}{4} \times y = 101\frac{1}{4}$ $y =$ _____

20. $p \times 3\frac{3}{8} = 67\frac{1}{2}$ $p =$ _____

Problem Solving
Solve.

21. Jenna is making wood shelves that are $16\frac{5}{8}$ inches long. If Jenna makes 6 shelves, how many inches of wood does she use?

_____

22. Juanita is making pies that use $1\frac{3}{4}$ pounds of cherries each. How many pounds of cherries does Juanita need to make 4 pies?

_____

Use with Grade 5, Chapter 14, Lesson 1, pages 322–325.

Name _____

Multiply a Mixed Number by a Fraction

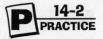

Multiply. Write each answer in simplest form.

1. $\frac{1}{5} \times 2\frac{1}{8} =$ _____

2. $\frac{4}{5} \times 3\frac{3}{4} =$ _____

3. $\frac{5}{6} \times 4\frac{2}{5} =$ _____

4. $7\frac{1}{2} \times \frac{1}{4} =$ _____

5. $6\frac{4}{9} \times \frac{2}{3} =$ _____

6. $\frac{6}{7} \times 10\frac{3}{4} =$ _____

7. $\frac{3}{4} \times 3\frac{1}{6} =$ _____

8. $\frac{5}{12} \times 8\frac{1}{10} =$ _____

9. $4\frac{3}{8} \times \frac{4}{9} =$ _____

10. $\frac{2}{5} \times 8\frac{1}{8} =$ _____

11. $5\frac{7}{10} \times \frac{3}{5} =$ _____

12. $\frac{5}{12} \times 9\frac{5}{6} =$ _____

13. $\frac{8}{9} \times 9\frac{3}{4} =$ _____

14. $3\frac{3}{7} \times \frac{14}{15} =$ _____

15. $1\frac{2}{3} \times \frac{9}{16} =$ _____

16. $\frac{3}{4} \times 20\frac{3}{4} =$ _____

17. $\frac{9}{10} \times 12\frac{5}{6} =$ _____

18. $12\frac{1}{2} \times \frac{4}{7} =$ _____

19. $\frac{2}{3} \times 4\frac{1}{4} \times \frac{1}{6} =$ _____

20. $12\frac{1}{2} \times \frac{3}{4} \times \frac{4}{5} =$ _____

Algebra Find the missing number.

21. $\frac{f}{5} \times 20 = 16$ $f =$ _____

22. $\frac{p}{3} \times 24 = 16$ $p =$ _____

23. $\frac{s}{5} \times 40 = 16$ $s =$ _____

24. $\frac{y}{9} \times 18 = 16$ $y =$ _____

25. $\frac{k}{6} \times 30 = 25$ $k =$ _____

26. $\frac{n}{8} \times 64 = 24$ $n =$ _____

Problem Solving
Solve.

27. Sandy runs at the rate of $5\frac{1}{2}$ miles per hour. Sandy runs for $\frac{3}{4}$ hour. How many miles does Sandy run?

_____

28. A pool is $8\frac{1}{2}$ feet deep. If the pool is only $\frac{1}{2}$ full, how deep is the water?

_____

Multiply Mixed Numbers

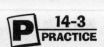

Multiply. Write each answer in simplest form.

1. $2\frac{1}{3} \times 5\frac{2}{5} =$ _____

2. $3\frac{1}{9} \times 1\frac{2}{3} =$ _____

3. $3\frac{3}{4} \times 4\frac{5}{6} =$ _____

4. $1\frac{3}{10} \times 3\frac{1}{6} =$ _____

5. $4\frac{1}{2} \times 6\frac{1}{3} =$ _____

6. $6\frac{7}{10} \times 1\frac{1}{4} =$ _____

7. $8\frac{5}{8} \times 7\frac{3}{5} =$ _____

8. $3\frac{1}{6} \times 9\frac{4}{5} =$ _____

9. $4\frac{2}{9} \times 3\frac{1}{6} =$ _____

10. $3\frac{2}{3} \times 2\frac{1}{2} =$ _____

11. $3\frac{3}{4} \times 2\frac{2}{5} =$ _____

12. $1\frac{1}{8} \times 1\frac{3}{10} =$ _____

13. $8\frac{5}{6} \times 2\frac{7}{9} =$ _____

14. $6\frac{1}{5} \times 3\frac{3}{4} =$ _____

15. $5\frac{3}{5} \times 6\frac{3}{7} =$ _____

16. $2\frac{7}{10} \times 4\frac{1}{9} =$ _____

17. $12\frac{1}{2} \times 7\frac{3}{5} =$ _____

18. $6\frac{3}{4} \times 8\frac{7}{8} =$ _____

19. $10\frac{9}{10} \times 1\frac{1}{3} =$ _____

20. $6\frac{7}{9} \times 3\frac{1}{4} =$ _____

21. $4\frac{3}{8} \times 17\frac{1}{2} =$ _____

22. $3\frac{4}{5} \times 4\frac{1}{3} \times 2 =$ _____

23. $\frac{1}{8} \times 8\frac{3}{4} \times 4 =$ _____

24. $\frac{2}{3} \times 1\frac{4}{5} \times 7\frac{3}{10} =$ _____

25. $\frac{1}{8} \times 10\frac{1}{3} \times 2\frac{1}{2} =$ _____

26. $1\frac{2}{3} \times 3\frac{2}{3} \times 6\frac{1}{4} =$ _____

27. $3\frac{3}{4} \times 2\frac{2}{5} \times 1\frac{5}{6} =$ _____

28. $5\frac{7}{8} \times 4\frac{3}{5} \times \frac{5}{6} =$ _____

29. $3\frac{5}{7} \times 7 \times 4\frac{5}{9} =$ _____

Problem Solving
Solve.

30. The Parks Department uses $1\frac{3}{4}$ gallons of paint for each picnic shelter. At the end of the first day, the workers had painted $2\frac{1}{2}$ shelters. How much paint had they used that day?

_____

31. While cleaning up around the picnic shelters, the workers filled $6\frac{1}{2}$ plastic bags with trash. If the average weight of a bag was $3\frac{3}{4}$ pounds, how many pounds of trash were collected?

_____

Problem Solving: Strategy
Make an Organized List

Make an organized list to solve.

1. Marsha spins both spinners and finds the product of the fractions. What products can Marsha make?

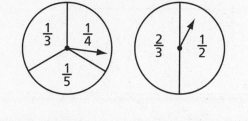

_____

2. David spins both spinners and finds the product of the fractions and the mixed numbers. What products can David make?

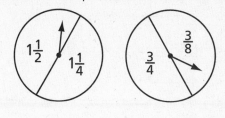

_____

3. Allie has square beads that are red, blue, and green. She has round beads that are yellow and white. If she chooses one color from each shape of beads, how many combinations of colors can she have?

_____

4. **Health** Ms. Dawson eats a fruit and a vegetable for lunch each day. She selects an apple, a banana, an orange, or a pear for her fruit. She chooses carrot sticks, celery sticks, or green-pepper slices for her vegetable. How many combinations of 1 fruit and 1 vegetable can she make?

_____

Mixed Strategy Review
Solve. Use any strategy.

5. **Time** Mario is making a new recipe for dinner. It takes $1\frac{1}{2}$ hours to prepare and $1\frac{1}{4}$ hours to bake. If he wants to serve dinner at 6:00 P.M., what time should Mario begin preparing the new recipe?

_____

Strategy: _____

6. Greta orders stickers that come with 12 sheets per package. Each sheet has 10 rows of stickers and each row has 8 stickers. How many stickers are in each package?

_____

Strategy: _____

7. Jamal, Katie, and Marion are standing in a line to buy tickets. In how many different ways can they order themselves in the line?

_____

Strategy: _____

8. **Write a problem** that you could make an organized list to solve. Share it with others.

_____

Name _____

Explore Dividing by Fractions

Use fraction models to find these quotients.

1. $3 \div \frac{1}{2} =$ _____

1
1
1

$\frac{1}{2}$

2. $2 \div \frac{1}{6} =$ _____

1
1

$\frac{1}{6}$

3. $1 \div \frac{1}{5} =$ _____

1

$\frac{1}{5}$

4. $2 \div \frac{1}{8} =$ _____

1
1

$\frac{1}{8}$

5. $3 \div \frac{1}{5} =$ _____

1
1
1

$\frac{1}{5}$

6. $2 \div \frac{1}{10} =$ _____

1
1

$\frac{1}{10}$

7. $3 \div \frac{1}{6} =$ _____

1
1
1

$\frac{1}{6}$

8. $4 \div \frac{1}{2} =$ _____

1
1
1

$\frac{1}{2}$

Use with Grade 5, Chapter 14, Lesson 5, pages 332–333.

Divide Fractions

Write the reciprocal of each number.

1. $\frac{2}{3}$ _____   **2.** $\frac{3}{5}$ _____   **3.** $\frac{1}{7}$ _____   **4.** $\frac{5}{6}$ _____

5. 3 _____   **6.** $\frac{7}{8}$ _____   **7.** $\frac{1}{4}$ _____   **8.** $\frac{11}{12}$ _____

9. $1\frac{3}{10}$ _____   **10.** $3\frac{1}{2}$ _____   **11.** $5\frac{4}{5}$ _____   **12.** $2\frac{5}{9}$ _____

Divide. Write each answer in simplest form.

13. $\frac{1}{3} \div \frac{1}{4} =$ _____   **14.** $\frac{1}{2} \div \frac{4}{5} =$ _____   **15.** $\frac{2}{3} \div 8 =$ _____

16. $4\frac{2}{3} \div \frac{2}{3} =$ _____   **17.** $\frac{5}{8} \div \frac{3}{4} =$ _____   **18.** $\frac{3}{4} \div 1\frac{1}{2} =$ _____

19. $\frac{5}{6} \div 5 =$ _____   **20.** $\frac{3}{5} \div \frac{4}{5} =$ _____   **21.** $\frac{3}{10} \div 9 =$ _____

22. $5\frac{5}{8} \div \frac{1}{4} =$ _____   **23.** $2\frac{4}{5} \div 7 =$ _____   **24.** $\frac{1}{3} \div \frac{2}{3} =$ _____

25. $6 \div \frac{3}{8} =$ _____   **26.** $9\frac{1}{6} \div 2\frac{1}{3} =$ _____   **27.** $3\frac{5}{12} \div \frac{1}{12} =$ _____

Complete.

28. $1\frac{1}{4} \div a = 5$ $a =$ _____   **29.** $4\frac{4}{5} \div a = 2\frac{2}{5}$ $a =$ _____

30. $3\frac{1}{2} \div n = 7$ $n =$ _____   **31.** $1\frac{1}{2} \div w = 2$ $w =$ _____

Problem Solving
Solve.

32. It takes $\frac{7}{8}$ inch of wire to make a small paper clip. How many small paper clips can be made from a piece of wire that is 14 inches long?

33. Lewis uses $\frac{1}{2}$ of a container to hold large paper clips. Each large paper clip takes up $\frac{1}{16}$ of the container. How many large paper clips are in the container?

_____   _____

Time

Complete.

1. 9 min = _____ s   **2.** 96 h = _____ d   **3.** 900 y = _____ centuries

4. 15 wk = _____ d   **5.** 12 h = _____ min   **6.** 730 d = _____ y

7. 7 decades = _____ y   **8.** 350 s = _____ min _____ s   **9.** 58 h = _____ d _____ h

10. 72 mo = _____ y   **11.** 6 d 9 h = _____ h   **12.** 60 d = _____ wk _____ d

Find each elapsed time.

13. 3:00 P.M. to 9:45 P.M.   **14.** 7:45 A.M. to 11:03 A.M.   **15.** 2:19 P.M. to 8:38 P.M.

_____   _____   _____

16. 11:12 P.M. to 4:05 A.M.   **17.** 8:40 A.M. to 2:56 P.M.   **18.** 7:32 P.M. to 2:26 A.M.

_____   _____   _____

Find each time.

19. 2 h 15 min after 1:30 P.M.   **20.** 5 h 26 min after 10:18 P.M.

_____   _____

21. 3 h 49 min after 6:45 A.M.   **22.** 8 h 57 min after 9:53 A.M.

_____   _____

Add or subtract.

23.   5 h 15 min   **24.**   6 h 10 min   **25.**   1 h 40 min
      + 3 h 35 min   − 3 h 20 min   + 3 h 45 min

Problem Solving
Solve.

26. A bus trip from New York City to Boston takes 4 hours 15 minutes. If you must be in Boston by 1:00 P.M., can you take a bus that leaves at 8:40 A.M.? Why or why not?

_____

_____

27. The ride from the bus station to downtown takes 35 minutes by taxicab. If you are meeting a friend at 2:20 P.M., what is the latest you should get into the taxicab?

_____

_____

Name _____

Nonstandard Measures

Use paperclips of the same size to measure the length.

1. _____

_____

2.

_____

3. red

_____

4.

_____

5.

_____

Use the width of your thumb to make each measurement.

6. the length of a sheet of paper _____

7. the width of a sheet of paper _____

8. the width of your shoe _____

9. the length of your desktop _____

10. the width of your desktop _____

Name _____

Customary Length

Measure the length of the pencil to the nearest:

1. inch _____   **2.** half inch _____   **3.** quarter inch _____   **4.** eighth inch _____

Choose an appropriate tool and unit to measure the length of each. Write *in.,*
ft, yd, or *mi.*

5. distance from Boston to Dallas

_____

6. height of a giraffe

_____

7. length of an aircraft carrier

_____

8. width of a computer diskette

_____

Complete.

9. 900 in. = _____ yd

10. 46 yd = _____ ft

11. 948 in. = _____ ft

12. 1,218 ft = _____ yd

13. 19 yd = _____ in.

14. 62 ft = _____ in.

15. 1,332 in. = _____ yd

16. 792 ft = _____ yd

17. 127 ft = _____ in.

18. 153 in. = _____ yd _____ in.

19. 26 ft = _____ yd _____ ft

20. 113 in. = _____ ft _____ in.

21. 263 in. = _____ yd _____ in.

22. 519 in. = _____ ft _____ in.

23. 178 ft = _____ yd _____ in.

Problem Solving
Solve.

24. A piece of red ribbon is $4\frac{1}{2}$ ft long. A
piece of blue ribbon is 1 yd long. How
many feet longer is the piece of red
ribbon than the piece of blue ribbon?
How many inches longer?

_____

25. A bookcase is 6 ft wide and a table is
30 in. wide. Will both fit along a wall
that is 3 yd long? Why or why not?

_____

Customary Capacity and Weight

Choose an appropriate unit to measure the capacity of each. Write *fl oz*, *c*, *pt*, *qt*, or *gal*.

1. drinking glass _____

2. kitchen sink _____

3. shampoo bottle _____

Choose an appropriate unit to measure the weight of each. Write *oz*, *lb*, or *T*.

4. bowling ball _____

5. compact disc _____

6. ocean liner _____

Complete.

7. 38 pt = _____ qt

8. 9 c = _____ fl oz

9. $4\frac{1}{2}$ T = _____ lb

10. 15 pt = _____ c

11. 64 oz = _____ lb

12. 32 qt = _____ gal

13. 21 fl oz = _____ c _____ fl oz

14. 2,450 lb = _____ T _____ lb

15. 26 qt = _____ gal _____ qt

16. 85 c = _____ pt _____ c

17. 6,500 lb = _____ T _____ lb

18. 19 pt = _____ qt _____ pt

Compare. Write >, <, or =.

19. 63 c $\bigcirc$ 129 pt

20. 5 lb $\bigcirc$ 50 oz

21. 164 c $\bigcirc$ 82 pt

22. 7,000 lb $\bigcirc$ 3 T

23. 65 gal $\bigcirc$ 256 qt

24. 60 fl oz $\bigcirc$ 10 c

25. 12 qt 1 pt $\bigcirc$ 25 pt

26. 55 oz $\bigcirc$ 3 lb 10 oz

27. 50 qt $\bigcirc$ 12 gal 1 qt

Problem Solving

Solve.

28. Shannon combines 3 quarts of cranberry juice with 3 pints of apple juice. Does Shannon now have at least one gallon of cranapple juice? Why or why not?

29. Mr. Hill's truck weighs $1\frac{1}{2}$ tons. His car weighs 1,600 pounds. Which vehicle weighs more? How much more?

Name _____

Problem Solving: Skill
Check for Reasonableness

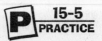

Is each estimate reasonable? Explain.

1. Sandra needs to buy a phone cord that will reach a distance of at least 12 yards. At the store, all of the packages are marked in feet. Sandra estimates that the package with 40 feet of cord will be enough. Is her estimate reasonable?

_____

_____

2. Kyle and Julie are watching a television program on weight lifting. A man is going to lift 210 pounds. Julie comments that he is going to lift 4,000 ounces. Is her estimate reasonable?

_____

_____

3. Ryan and Tyler are going to the pet shop to buy 12 cans of dog food. They are trying to decide whether they should take their wagon to help carry the dog food home. The cans weigh 15 ounces each. They estimate that the dog food will weigh 10 pounds. Is the estimate reasonable?

_____

_____

4. Nicole is trying out a new recipe. The recipe calls for 4 pints of broth. Nicole has only a 1-cup measuring cup. She estimates that she will need 16 cups of broth. Is her estimate reasonable?

_____

_____

5. A scoutmaster needs 10 feet of nylon cord to teach his scouts how to tie different types of knots. The cord is sold in yards. He estimates that he should buy 30 yards of the nylon cord. Is his estimate reasonable?

_____

_____

Use with Grade 5, Chapter 15, Lesson 5, pages 366–367.

Explore Metric Length

Measure to the nearest centimeter and millimeter.

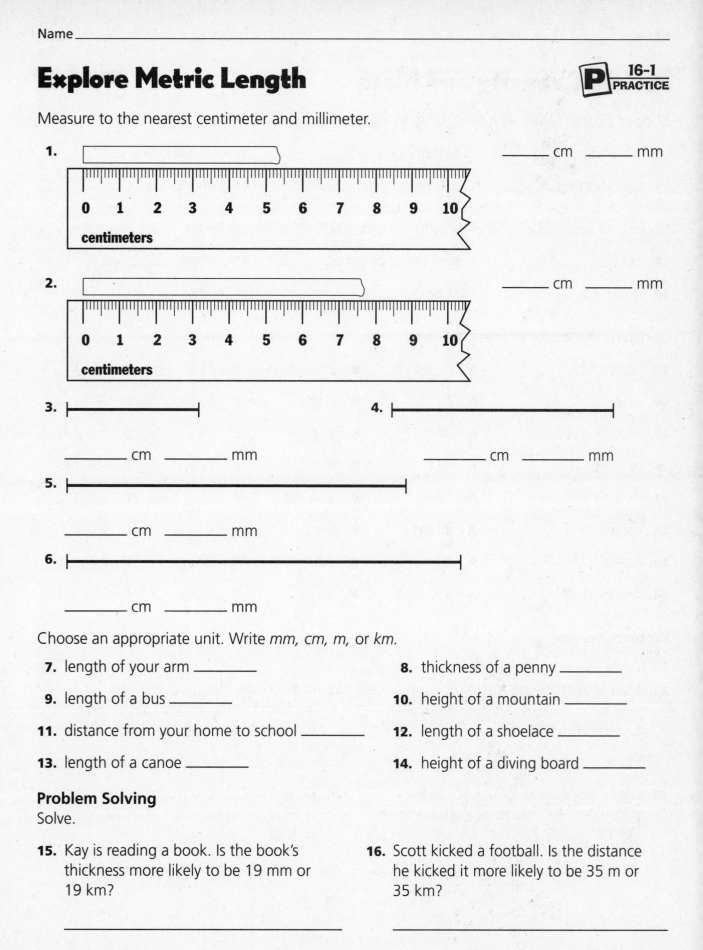

1. _____ cm _____ mm

2. _____ cm _____ mm

3. _____ cm _____ mm

4. _____ cm _____ mm

5. _____ cm _____ mm

6. _____ cm _____ mm

Choose an appropriate unit. Write *mm, cm, m,* or *km.*

7. length of your arm _____

8. thickness of a penny _____

9. length of a bus _____

10. height of a mountain _____

11. distance from your home to school _____

12. length of a shoelace _____

13. length of a canoe _____

14. height of a diving board _____

Problem Solving
Solve.

15. Kay is reading a book. Is the book's thickness more likely to be 19 mm or 19 km?

_____

16. Scott kicked a football. Is the distance he kicked it more likely to be 35 m or 35 km?

_____

Metric Capacity and Mass

P

Choose an appropriate unit of capacity to measure each. Write *mL* or *L*.

1. water glass _____

2. bath tub _____

3. ice cream cone _____

4. baby bottle _____

5. watering can _____

6. gasoline can _____

Choose an appropriate unit of mass to measure each. Write *mg*, *g*, or *kg*.

7. cat _____

8. sheet of paper _____

9. sandwich _____

10. quarter _____

11. brick _____

12. feather _____

Circle an appropriate estimate for each.

13. coffee mug	**A** 25 mL	**B** 250 mL	**C** 25 L	**D** 250 L
14. sink	**A** 2 L	**B** 2 mL	**C** 200 mL	**D** 20 L
15. medicine dropper	**A** 30 mL	**B** 3 mL	**C** 30 L	**D** 3 L
16. water bucket	**A** 4 L	**B** 40 L	**C** 400 L	**D** 400 mL
17. dog	**A** 12 kg	**B** 120 mg	**C** 12 g	**D** 120 g
18. nickel	**A** 50 mg	**B** 5 kg	**C** 50 kg	**D** 5 g
19. apple	**A** 2 g	**B** 20 mg	**C** 200 g	**D** 2 kg
20. box of cereal	**A** 350 g	**B** 35 g	**C** 35 kg	**D** 350 mg

Problem Solving
Solve.

21. Marc was telling his friends about his new baby sister. Is her mass more likely to be 40 mg or 4 kg?

_____

22. Kate bought her mother a vase for a present. Is the volume of the vase more likely to be 800 mL or 80 L?

_____

23. Lila's class drank a lot of juice at the party. Was the volume of the juice more likely 75 mL or 75 L?

_____

24. Gavin likes to hold his pet cat, Shadow. Is Shadow's mass more likely to be 6 kg or 6 g?

_____

Use with Grade 5, Chapter 16, Lesson 2, pages 374–376.

Algebra: Metric Conversions

Complete.

1. 26 cm = _____ mm

2. 745 cm = _____ m

3. 8.4 km = _____ m

4. 350 mL = _____ L

5. 93 cL = _____ mL

6. 13.5 L = _____ mL

7. 65 kg = _____ g

8. 16 g = _____ kg

9. 52 mg = _____ g

10. 3.07 L = _____ mL

11. 0.6 m = _____ cm

12. 44.2 g = _____ kg

13. 62 mL = _____ cL

14. 6,400 L = _____ mL

15. 250 mm = _____ cm

16. 4,500 mm = _____ m

17. 7.2 kg = _____ g

18. 800 cm = _____ mm

Find each sum.

19. 650 g + 2 kg + 195 g = _____ kg

20. 36.7 g + 24.8 g + 513 mg = _____ g

21. 580 m + 1.2 km + 7 km = _____ km

22. 53 cm + 124 cm + 3.4 m = _____ m

23. 3.4 L + 16 mL + 297 mL = _____ L

24. 6.8 cL + 156 mL + 94 cL = _____ cL

Compare. Write >, <, or =.

25. 520.8 cm ◯ 5,208 mm

26. 320 m ◯ 3.2 km

27. 295 cm ◯ 29.5 m

28. 6.34 m ◯ 63.4 cm

29. 2,000 mL ◯ 20 cL

30. 4.027 L ◯ 4,027 mL

31. 129 mL ◯ 12.9 L

32. 56.8 cL ◯ 568 mL

33. 4,300 g ◯ 0.43 kg

34. 0.9 g ◯ 900 mg

35. 2.45 kg ◯ 245 g

36. 0.384 g ◯ 3,840 mg

Problem Solving

Solve.

37. Jacob has 0.5 L of milk to use in two recipes. Each recipe uses 300 mL. Does he have enough? Explain.

_____

_____

38. When completed, a tunnel will be 1.3 km long. The workers have already completed 825 m of the tunnel. How many meters more of the tunnel remains to be built?

_____

Problem Solving: Strategy
Draw a Diagram

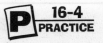

Draw a diagram to solve.

1. For a concert, Ron must set the speakers for a sound system every 10 yards around the walls of a square room. Speakers are not set up in corners of the room. The room is 60 yards long. How many speakers will Ron set up?

_____

2. Katya makes a 4-by-4 grid. She writes the numbers 0 through 15 in order on the grid, starting with the top left square. What are the four numbers in the right column of the grid?

_____

3. Pine cones are evenly spaced on a circular wreath. The third pine cone is opposite the ninth pine cone. How many pine cones are on the wreath?

_____

4. Jason is building a dog run that is 24 feet by 18 feet. He is setting a fence post every 6 feet and one at each corner. How many posts will he need in all?

_____

Mixed Strategy Review

Solve. Use any strategy.

5. Tami, Evan, and Scott each prefer a different type of music. They listen to rock, rap, and country. Tami does not like country. Evan does not like country or rap. Which type of music does each person like best?

_____

Strategy: _____

6. **Social Studies** The writer F. Scott Fitzgerald was born in St. Paul, Minnesota in 1896. The city of his birth was first called Pig's Eye when it was established 56 years earlier. The name of the city was changed to St. Paul one year after it was established. What year was the city named St. Paul?

_____

Strategy: _____

7. Renaldo paid for a roll and a large glass of juice with 9 coins. If the total cost for both items was $2.52, which coins did Renaldo use to pay the bill?

_____

Strategy: _____

8. **Write a problem** that you could draw a diagram to solve. Share it with others.

_____

_____

Use with Grade 5, Chapter 16, Lesson 4, pages 382–383.

Name _____

Integers and the Number Line

Write an integer to represent each situation.

1. spent $15 _____

2. 11 degrees colder than 0°F _____

3. 8-yard gain in football _____

4. deposit of $25 into bank account _____

5. 10 feet below sea level _____

6. 3-centimeter increase in height _____

7. withdrawal of $50 from bank account _____

8. received $5 allowance _____

9. speed increase of 15 mph _____

10. 30 seconds before liftoff _____

Describe a situation that can be represented by the integer.

11. ⁻17 _____

12. ⁺$27 _____

13. ⁺45 _____

14. ⁻9 _____

Compare. Write < or >. You may use a number line.

15. ⁻2 ◯ 4 **16.** 3 ◯ ⁻7 **17.** ⁻6 ◯ ⁻9 **18.** ⁻5 ◯ 1

19. 6 ◯ ⁻8 **20.** ⁻4 ◯ 0 **21.** ⁻3 ◯ ⁻10 **22.** 6 ◯ ⁻6

23. ⁻12 ◯ ⁻10 **24.** 13 ◯ ⁻17 **25.** 0 ◯ ⁻17 **26.** ⁻14 ◯ 21

Problem Solving
Solve.

27. The low temperature on Saturday was ⁻5°F. The low temperature on Sunday was ⁻9°F. Which day was colder?

28. On one play a football team moved the ball ⁻6 yards. On the next play, they moved the ball exactly the opposite. Did they gain or lose yards on the second play? How many yards?

_____

_____

Temperature

Estimate each temperature in °F and °C.

1. comfortable room temperature

_____

2. cup of hot chocolate

_____

3. warm day

_____

4. glass of cold milk

_____

5. icy day

_____

6. body temperature

_____

Estimate each Fahrenheit temperature. Show your work.

7. 3°C ≈ _____ °F

_____

8. 90°C ≈ _____ °F

_____

9. 32°C ≈ _____ °F

_____

10. 66°C ≈ _____ °F

_____

11. 85°C ≈ _____ °F

_____

12. 13°C ≈ _____ °F

_____

13. 20°C ≈ _____ °F

_____

14. 49°C ≈ _____ °F

_____

15. 34°C ≈ _____ °F

_____

16. 23°C ≈ _____ °F

_____

17. 51°C ≈ _____ °F

_____

18. 27°C ≈ _____ °F

_____

Problem Solving
Solve.

19. A weather forecast predicts that temperatures will fall from a high of 2°C and there will be precipitation. What form do you think the precipitation will take? Explain.

_____

_____

_____

20. Yesterday's low temperature was 18°C. The high temperature was 26°C. What was the temperature range in degrees Fahrenheit? Explain.

_____

_____

_____

Name_____

Explore Addition and Subtraction Expressions

1. A scout troop planted bushes in a park. They planted 3 fewer lilac bushes than rose bushes. How many lilac bushes did they plant?

   Draw □ to show the number of lilac bushes planted when certain numbers of rose bushes are planted. Let each □ represent 1 lilac bush.

   5 rose bushes   6 rose bushes   7 rose bushes   8 rose bushes   9 rose bushes

   Use your drawing to complete the table.

Number of Rose Bushes	5	6	7	8	9
Number of Lilac Bushes					

2. Write an expression for the relationship between the number of lilac bushes and rose bushes. Use the variable, r, to represent the number of rose bushes.

   _____

3. Suppose the troop planted 12 rose bushes. How many lilac bushes did they plant? Evaluate the expression you wrote for $r = 12$. Show your work.

   _____

Write an expression for each situation.

4. John worked x hours planting bushes. Kim worked 2 more hours than John. How many hours did Kim work?

   _____

5. A rose bush costs x dollars. A lilac bush costs $2.50 less than a rose bush. How much does a lilac bush cost, in dollars?

   _____

6. The lilac bush is x feet tall now. By next year, it should be $2\frac{1}{2}$ feet taller. How tall will the lilac bush be then, in feet?

   _____

7. The troop has x bushes to plant. They have already planted 8 bushes. How many bushes do they still have to plant?

   _____

Problem Solving
Solve.

8. This year the troop planted 15 more bushes than last year. Write an expression for the number planted this year. Let y represent the number planted last year.

   _____

9. Last year the troop planted 12 bushes. Evaluate the expression you wrote in problem 8 to find how many bushes they planted this year.

   _____

Name _____

Explore Multiplication and Division Expressions

1. At a nature preserve, visitors are divided into groups of 3 to go on bird watches. How many groups are there?

Draw □ to show the number of groups when there are certain numbers of visitors. Let each □ represent one group.

9 visitors   12 visitors   15 visitors   18 visitors   21 visitors

Use your drawings to complete the table.

Number of visitors	9	12	15	18	21
Number of groups					

2. Write an algebraic expression for the relationship between the number of groups and the number of visitors. Use the variable v to represent the number of visitors.

_____

3. Suppose there were 36 visitors. How many groups are there? Evaluate the expression you wrote for $v = 36$. Show your work.

_____

Write an expression for each situation.

4. Every year the school's Science Club builds 2 bird feeders. How many bird feeders will the club build in x years?

_____

5. A hiker walks 3 miles per hour. How many hours does it take the hiker to walk n miles?

_____

6. Volunteers clean up the nature trail at the nature preserve. The trail is x miles long. Each volunteer is to clean up 0.3 mile of the trail. How many volunteers are needed to clean up the whole trail?

_____

7. Each day a naturalist puts $1\frac{1}{2}$ pounds of birdseed in each feeder. There are x feeders. How many pounds of seed are needed for one day?

_____

94

Name_____

Order of Operations

Evaluate. Use the order of operations.

1. $44 + 7 \times 3$ _____

2. $48 \div (8 - 2)$ _____

3. $(3 + 4) \times 8 \div 2$ _____

4. $18 + 12 \div 2 + 3$ _____

5. $4^2 \times 2 - 10$ _____

6. $(3.6 \div 3) + (8 \times 5)$ _____

7. $(3 + 2) \times 3^2$ _____

8. $24 \div 6 \times 3 + 52$ _____

9. $(2\frac{1}{2} \times 2) - (3 \times \frac{1}{3})$ _____

10. $96 \div (3 \times 4) \div 2$ _____

11. $100 - 8^2 + 4 \div 4$ _____

12. $(200 - 50) \div (12 - 9)$ _____

13. $47 + 3 \times 11 - 36 \div 3$ _____

14. $(3.7 + 6.3) \times (7.2 - 3.1)$ _____

15. $50 - (^-4 + 1)^2 \div 9$ _____

16. $6^2 - 9 \times 4 + (\frac{1}{3})^2$ _____

Evaluate the expression for the value given.

17. $37 + 2p$ for $p = 6$ _____

18. $\frac{45}{a} - 2$ for $a = 5$ _____

19. $7 + s - 12$ for $s = 23$ _____

20. $3c - 15$ for $c = 21$ _____

21. $\frac{m}{4} + 3.9$ for $m = 40$ _____

22. $29 - h + 13$ for $h = 8$ _____

23. $(18 + 9) \div 3d$ for $d = 3$ _____

24. $8 + 24 \div 4k$ for $k = 2$ _____

25. $\frac{e}{3} - 2 \times 3$ for $e = 41.4$ _____

26. $(z^2 + 7) - 5 \times 10$ for $z = 9$ _____

27. $4d + 30$ for $d = 10$ _____

28. $\frac{k}{3} - 6.4$ for $k = 12$ _____

Problem Solving

Solve.

29. Tickets to the school play cost $4.50 for adults and $2.00 for students. If 255 adults and 382 students attended the play, write an expression that shows the total amount of money made on ticket sales. Then simplify the expression.

_____

_____

30. At the school play, popcorn costs $1 and juice costs $0.75. Suppose 235 people buy popcorn and 140 people buy juice. Write an expression that shows the total amount of money made by selling refreshments. Then simplify the expression.

_____

_____

Name _____

Functions

PRACTICE

Write an equation to describe each situation.
Tell what each variable represents.

1. Marie is sending some paperback books to her cousin. Each book weighs 4 ounces.
   She is mailing them in a box that weighs 6 ounces.

   Variables: _____

   _____

   Equation: _____

2. A mailing service charges $3 plus $2 a pound to wrap and send a package.

   Variables: _____

   _____

   Equation: _____

Complete the table. Then write an equation to describe the situation.
Tell what each variable represents.

3. **Cost of Ordering Puzzles from a Catalog**
   Cost of Each Puzzle: $7.50

Number of Puzzles	1	2	3	4	5
Total Cost	$12.50	$20.00	$27.50		

   Variables: _____

   _____

   Equation: _____

Use data from the information below to solve problems 4–5.

It takes Beth 20 minutes to drive to and from a mailing service and
2 minutes to fill out a mailing label and have each package weighed.

4. Write an equation to describe the situation. Tell what the variables represent.

5. How long will it take Beth to mail 3 packages? Use the equation you wrote to solve the problem.

© Macmillan/McGraw-Hill. All rights reserved.

Use with Grade 5, Chapter 17, Lesson 4, pages 408–410.

Name _____

Graphing a Function

Write the coordinates for each point.

1. A _____   2. I _____   3. F _____

4. L _____   5. G _____   6. D _____

Name the point for the ordered pair.

7. (8, 5) ____   8. (4, 4) ____   9. (2, 7) ____

10. (7, 0) ____   11. (3, 9) ____   12. (9, 9) ____

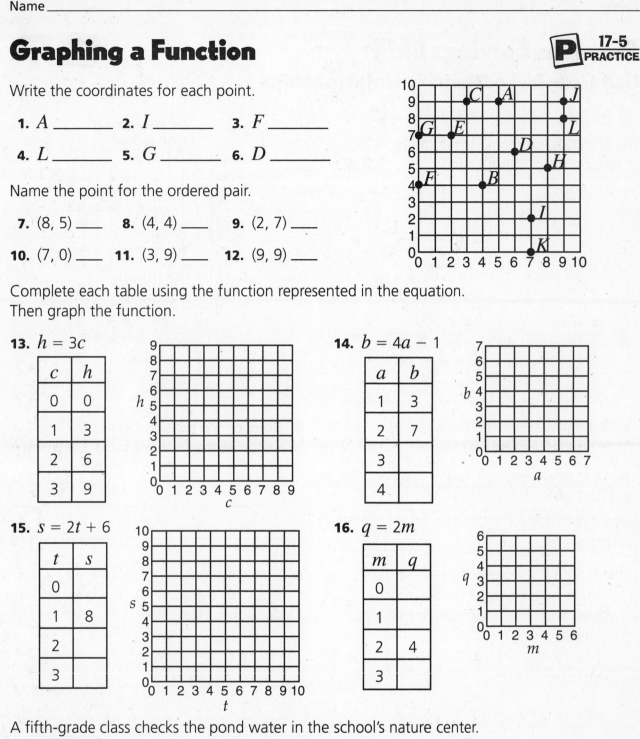

Complete each table using the function represented in the equation.
Then graph the function.

13. $h = 3c$

c	h
0	0
1	3
2	6
3	9

14. $b = 4a - 1$

a	b
1	3
2	7
3	
4	

15. $s = 2t + 6$

t	s
0	
1	8
2	
3	

16. $q = 2m$

m	q
0	
1	
2	4
3	

A fifth-grade class checks the pond water in the school's nature center.
Each day they collect some 4-ounce samples of water and one 8-ounce
sample of water.

17. Write an equation that describes the relationship between the total ounces of water collected, w, and the number of 4-ounce samples, s.

_____

18. What is the total amount of water that will be collected if students collect three 4-ounce samples?

_____

Name _____

Problem Solving: Skill
Use Graphs to Identify Relationships

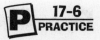

Use data from the graph for problems 1–7.

The graph shows the air temperature in Jackson for a 24-hour period.

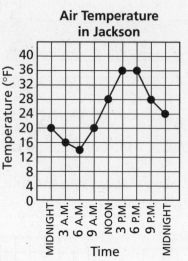

Air Temperature in Jackson

1. Describe how the temperature was changing from midnight to 6 A.M.

_____

2. During what time period was the temperature constant? Explain.

_____

_____

_____

3. Describe the relationship in the temperatures from 6 A.M. to 3 P.M. How does the graph show this change?

_____

_____

_____

4. By about how many degrees did the temperature change from 9 A.M. to 3 P.M.? Did the temperature increase or decrease?

_____

_____

5. During what part of the 24-hour period shown in the graph did the temperature decrease in Jackson?

_____

_____

_____

6. Between what times did the temperature fall about 2°F?

_____

7. What was the temperature difference from midnight to noon? From midnight to midnight?

_____

Use with Grade 5, Chapter 17, Lesson 6, pages 416–417.

Name _____

Explore Addition Equations

Each sheet of paper represents one side of the equation. The counters represent the numbers and the cups stand for the variables. Write the equation represented by each model. Then solve it.

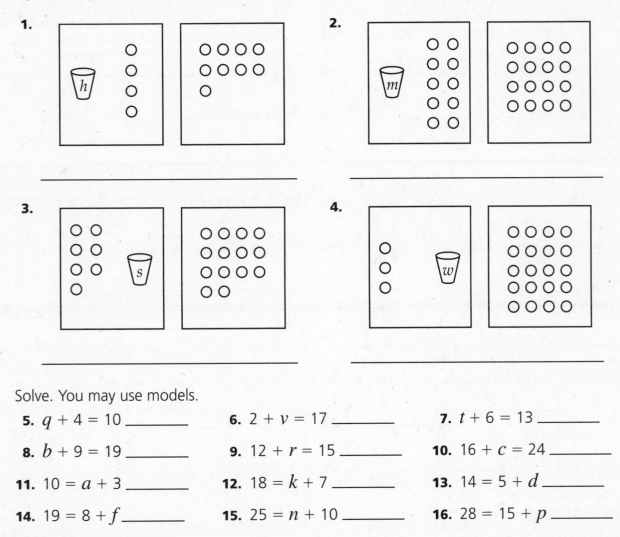

1. _____

2. _____

3. _____

4. _____

Solve. You may use models.

5. $q + 4 = 10$ _____

6. $2 + v = 17$ _____

7. $t + 6 = 13$ _____

8. $b + 9 = 19$ _____

9. $12 + r = 15$ _____

10. $16 + c = 24$ _____

11. $10 = a + 3$ _____

12. $18 = k + 7$ _____

13. $14 = 5 + d$ _____

14. $19 = 8 + f$ _____

15. $25 = n + 10$ _____

16. $28 = 15 + p$ _____

Use the data in the information below to solve problems 17 and 18.

A box contains 4 pounds of sculpting clay and some painting supplies. The contents of the box weighs 7 pounds in all.

17. Draw a model to represent the situation.

18. Write an equation represented by the model you drew. Solve it to find the weight of the painting supplies.

_____

Name _____

Addition Equations

Solve each equation. Check your answer.

1. $a + 8 = 23$ _____

2. $s + 9 = 26$ _____

3. $f + 36 = 58$ _____

4. $z + 16 = 59$ _____

5. $v + 14 = 162$ _____

6. $h + 2.7 = 3.8$ _____

7. $k + 60 = 84$ _____

8. $t + 30 = 94$ _____

9. $r + \frac{3}{4} = 17$ _____

10. $96 = d + 78$ _____

11. $s + 14.9 = 31.6$ _____

12. $100 = c + 42$ _____

13. $4.5 = e + 0.4$ _____

14. $z + 2\frac{1}{2} = 6\frac{3}{4}$ _____

15. $529 = g + 300$ _____

16. $c + 200 = 473$ _____

17. $w + 356 = 500$ _____

18. $p + \frac{2}{3} = 7$ _____

19. $e + 211 = 481.6$ _____

20. $923 = y + 127$ _____

21. $h + 41.8 = 48.2$ _____

22. $3\frac{1}{4} = q + 1\frac{5}{8}$ _____

23. $m + 32.7 = 75$ _____

24. $d + 4\frac{9}{16} = 12\frac{1}{8}$ _____

Problem Solving
Solve.

25. The high temperature one day in Washington, D.C., was 40°F. That was 14°F greater than the low temperature. Write an addition equation to describe the situation. Use t to represent the low temperature. Then solve the equation.

_____

_____

26. One year Chicago, IL, received 39.2 inches of snow. That was 9.8 inches more than the previous year. Write an addition equation to describe the situation. Solve it to find last year's snowfall in inches, s.

_____

_____

Use with Grade 5, Chapter 18, Lesson 2, pages 424–425.

Subtraction Equations

Solve each equation. Check your answer.

1. $n - 14 = 92$ _____

2. $d - 3.5 = 6$ _____

3. $b - 7\frac{5}{6} = 3\frac{2}{3}$ _____

4. $r - 41 = 49$ _____

5. $y - 8.6 = 14.2$ _____

6. $s - 4\frac{1}{3} = \frac{2}{3}$ _____

7. $19 = d - 62$ _____

8. $m - 7.6 = 8.9$ _____

9. $28.3 = w - 7.7$ _____

10. $46.3 = z - 7.6$ _____

11. $f - 3\frac{1}{3} = 12\frac{1}{4}$ _____

12. $2\frac{1}{8} = t - 1\frac{1}{8}$ _____

13. $a - \frac{7}{10} = 4\frac{3}{5}$ _____

14. $v - 0.9 = 25.2$ _____

15. $96.5 = i - 3.6$ _____

16. $30 - a = 21$ _____

17. $16 - v = 4$ _____

18. $3\frac{11}{12} = n - 1\frac{7}{12}$ _____

19. $e - 9.7 = 23.4$ _____

20. $9.7 - b = 8.9$ _____

21. $6\frac{1}{10} - a = 3\frac{1}{10}$ _____

22. $c - 0 = 4\frac{5}{9}$ _____

23. $298.7 = i - 1.6$ _____

24. $17\frac{3}{8} = r - 4\frac{2}{3}$ _____

Problem Solving
Solve.

25. Leah started with *d* dollars. After Leah spent $19, she had $13 left. Write a subtraction equation to represent this situation. Then solve the equation to find the amount of money Leah started with.

_____

_____

26. A chapter has 45 pages. Larry has read *n* pages, and has 8 pages left. Write a subtraction equation to represent this situation. Then solve the equation to find the number of pages Larry has left to read.

_____

_____

Multiplication and Division Equations

Solve each equation. Check your answer.

1. $7w = 28$ _____

2. $q \div 6 = 108$ _____

3. $d \times 20 = 180$ _____

4. $\frac{a}{9} = 7.2$ _____

5. $e \times 4 = 276$ _____

6. $15y = 48$ _____

7. $k \div 40 = 8$ _____

8. $0.4 \times p = 16$ _____

9. $\frac{j}{52} = 13$ _____

10. $s \div 12 = 60$ _____

11. $30h = 15$ _____

12. $w \div 0.8 = 64$ _____

13. $b \div \frac{5}{8} = 8$ _____

14. $2.4c = 120$ _____

15. $1\frac{1}{2} \times n = 15$ _____

16. $\frac{s}{0.7} = 21$ _____

17. $\frac{x}{1.2} = 1.2$ _____

18. $f \times 32 = 6.4$ _____

19. $0.6t = 60$ _____

20. $w \div \frac{1}{4} = 24$ _____

21. $\frac{z}{60} = 20$ _____

22. $\frac{3}{5}b = 12$ _____

23. $4.1 \times v = 20.5$ _____

24. $w \div 1.6 = 8$ _____

25. $r \times 8.6 = 21.5$ _____

26. $0.5x = 4.2$ _____

27. $9.5 \div s = 0.95$ _____

28. $d \div \frac{1}{3} = 225$ _____

29. $2\frac{1}{2} \times b = 50$ _____

30. $\frac{2}{3}a = 24$ _____

Problem Solving
Solve.

31. The Martinez family paid $37.50 for 5 movie passes. Write a multiplication equation to describe the situation. Solve it to find the cost in dollars, c, of each movie pass.

_____

32. Three friends split the cost of a gift equally. Each paid $4.19. Write a division equation to describe the situation. Solve it to find the total cost in dollars, t, of the gift.

_____

Use with Grade 5, Chapter 18, Lesson 4, pages 428–430.

Name _____

Problem Solving: Strategy
Make a Graph

Solve. Use a graph.

1. The average surface temperature on Venus is 870°F. Use the equation $F = 1.8°C + 32$ to estimate the average surface temperature on Venus in degrees Celsius. Graph this equation.

   _____

2. The surface temperature on Venus can reach 900°F. Use the graph from problem 1 to find this surface temperature in degrees Celsius.

   _____

3. A botanist finds that the height of a tree, in centimeters, is related to its age, in weeks, by the equation $y = 3x + 1$. Graph this equation.

   _____

4. If a tree measures to be 16 centimeters, how many weeks old is the tree? If the tree is 4 weeks old, how tall is the tree? Use the graph from problem 3 to solve.

   _____

Mixed Strategy Review

Solve. Use any strategy.

5. **Career** A jeweler has 3 necklaces and 7 pairs of earrings. How many different combinations of necklaces and earrings can he put together in a display?

   _____

   Strategy: _____

6. Ana Maria gave 5 flowers to her sister and twice as many to her mother. If she has a half dozen flowers left, how many flowers did she have to start with?

   _____

   Strategy: _____

7. **Literature** Jeremy has four times as many books by K.A. Applegate as by Lynne Reid Banks. If he has 30 books by the two authors, how many does he have by each one?

   _____

   Strategy: _____

8. **Create a problem** in which you must use a graph to solve the problem. Share it with others.

   _____

   _____

   _____

Name _____

Two-Step Equations

Solve.

1. $2c + 9 = 35$ _____

2. $\frac{v}{5} - 13 = 37$ _____

3. $7t - 3 = 60$ _____

4. $\frac{h}{4} + 7 = 12$ _____

5. $\frac{w}{20} - 1 = 15$ _____

6. $24 + 3a = 36$ _____

7. $73 = 12s - 23$ _____

8. $11y - 14 = 173$ _____

9. $8 + 1.3d = 36.6$ _____

10. $23 + \frac{b}{10} = 26$ _____

11. $18 = \frac{y}{2.5} - 2$ _____

12. $10 = \frac{3}{4}e + 7$ _____

13. $16m - 2 = 46$ _____

14. $\frac{s}{18} - \frac{1}{2} = 3\frac{1}{2}$ _____

15. $20 = \frac{z}{5} + 16$ _____

16. $5 = \frac{a}{3.1} - 1$ _____

17. $4 + \frac{p}{15} = 11$ _____

18. $\frac{5}{6} = \frac{2}{3}d - \frac{1}{6}$ _____

19. $25y + 7 = 157$ _____

20. $\frac{b}{12} - 2 = 10$ _____

21. $\frac{f}{0.4} + 1.6 = 9.6$ _____

22. $15 = 7.5 + 3u$ _____

23. $1\frac{5}{12} = \frac{3}{4} + 2w$ _____

24. $30t - 120 = 180$ _____

Problem Solving
Solve.

25. A large pizza with cheese costs $8.00. Each additional topping costs $0.75. Glenn ordered a large pizza with toppings that cost $10.25. Write a two-step equation to represent this situation. Solve it to find the number of additional toppings, t, on the pizza.

_____

_____

26. With a $2-off coupon, 4 sandwiches at a restaurant cost $5.16. Write a two-step equation to represent this situation. Solve it to find the original cost in dollars, c, of each sandwich.

_____

_____

Use with Grade 5, Chapter 18, Lesson 6, pages 434–435.

Basic Geometric Ideas

Identify each figure. Then name it using symbols.

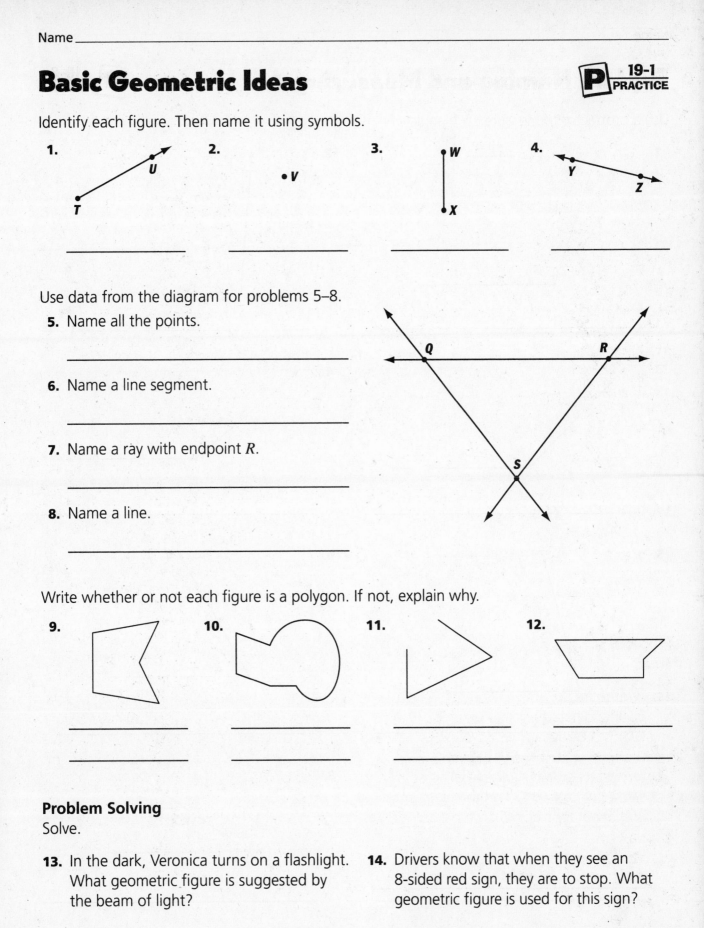

1.

2.

3.

4.

Use data from the diagram for problems 5–8.

5. Name all the points.

_____

6. Name a line segment.

_____

7. Name a ray with endpoint *R*.

_____

8. Name a line.

_____

Write whether or not each figure is a polygon. If not, explain why.

9.

10.

11.

12.

_____   _____   _____   _____

Problem Solving
Solve.

13. In the dark, Veronica turns on a flashlight. What geometric figure is suggested by the beam of light?

_____

14. Drivers know that when they see an 8-sided red sign, they are to stop. What geometric figure is used for this sign?

_____

Explore Naming and Measuring Angles

Use a protractor to measure each angle.

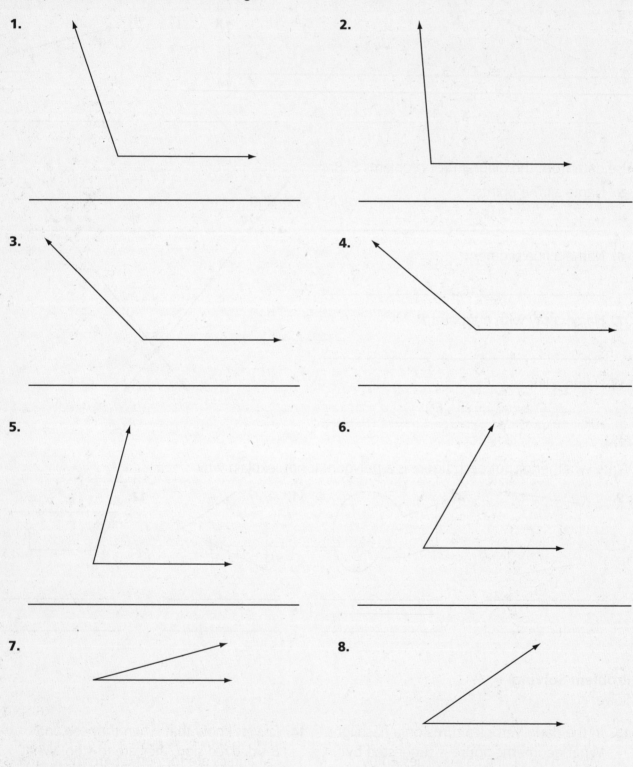

1.

2.

3.

4.

5.

6.

7.

8.

Use with Grade 5, Chapter 19, Lesson 2, pages 454–455.

Name _____

Classify Angles and Line Relationships

Use a protractor to measure each angle. Classify the angle as acute, right, or obtuse.

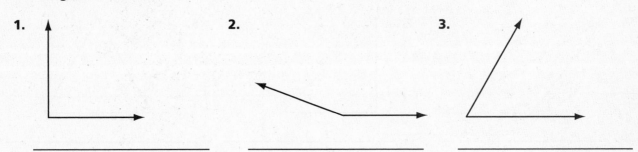

1.
_____

2.
_____

3.
_____

Name the pair of lines as intersecting, parallel, or perpendicular.

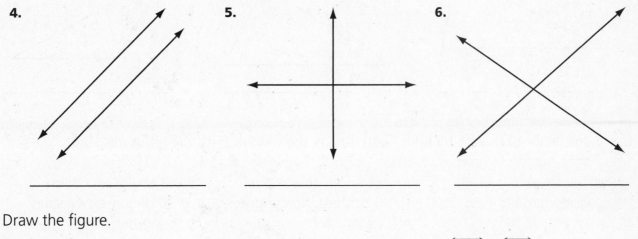

4.
_____

5.
_____

6.
_____

Draw the figure.

7. right angle *MNO*

8. obtuse angle *DEF*

9. $\overleftrightarrow{GH} \parallel \overleftrightarrow{KL}$

Problem Solving
Solve.

10. First Avenue and Main Street cross each other, making angles of 75° and 105°. What type of lines are suggested by these streets?

_____

11. First Avenue and Park Drive are perpendicular to each other. What type of angles are formed where these streets meet?

_____

Triangles

Classify each triangle as equilateral, isosceles, or scalene
and right, acute, or obtuse.

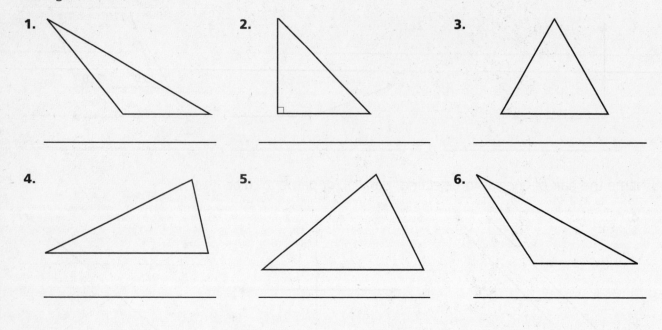

1. _____

2. _____

3. _____

4. _____

5. _____

6. _____

Use a protractor to draw a triangle. Then classify the triangle with the given measure.

7. all angles equal, all
sides congruent

8. one angle greater than
90 degrees, no congruent
sides

9. all angles less than
90 degrees, two sides
congruent

_____

_____

_____

Problem Solving
Solve.

10. Tyler draws a triangle with a 35° angle
and an 85° angle. What is the measure
of the third angle?

11. Amber draws an obtuse, isosceles
triangle with a 110° angle. What are the
measures of the other two angles?

_____

_____

Use with Grade 5, Chapter 19, Lesson 4, pages 458–460.

Name _____

Quadrilaterals

Classify the quadrilateral in as many ways as possible.
Write the sum of the measure of the angles for each one.

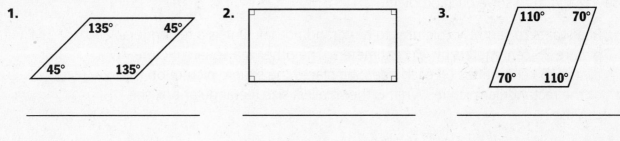

1. 135° 45° / 45° 135°

2.

3. 110° 70° / 70° 110°

_____   _____   _____

_____   _____   _____

_____

Write *true* or *false*. Explain your reasoning.

4. All squares are rhombuses.

_____

_____

5. All trapezoids have exactly one pair of congruent sides.

_____

6. All rhombuses are parallelograms.

_____

_____

Draw and classify each quadrilateral described for problems 7–9.

7. two pairs of congruent sides

8. two pairs of parallel sides

9. no right angles

_____   _____   _____

Problem Solving
Solve.

10. Lee drew a quadrilateral with three angles that measure 120 degrees, 110 degrees, and 70 degrees. What is the measure of the fourth angle?

_____

11. Robert drew a parallelogram with two 55-degree angles. What are the measures of the other two angles?

_____

Problem Solving: Skill
Use a Diagram

Use a diagram to solve each problem.

1. Rita wants to send two pictures to her grandmother. One is a rectangular picture 25 centimeters by 40 centimeters. The other is a square picture that is 35 centimeters on each side. She places the square picture on top of the rectangular picture. What is the smallest size rectangular box she can use?

   _____

2. Manuel made two mosaics. He wants to send them to his father. One is a rectangular mosaic 50 centimeters by 30 centimeters. The other is a circular mosaic with a 40-centimeter diameter. He places the circular mosaic on top of the rectangular mosaic to send them. What is the smallest size rectangular box he can use?

   _____

3. Lily is sending a picture frame to her cousin. The frame measures 12 inches by 18 inches. The shipping store only sells square boxes. What size must she buy?

   _____

4. Austin has two stained glass pictures that he wants to send to his grandparents. One is a rectangular picture that measures 30 centimeters by 20 centimeters. The other is a circular picture with a diameter of 24 centimeters. He puts the circular picture on top of the rectangular picture to send them. What is the smallest size rectangular box he can use?

   _____

5. Max has two paintings to ship. One is a rectangular painting that is 70 centimeters by 90 centimeters. The other is shaped like an equilateral triangle with sides that are 80 centimeters long. He stacks the triangular painting on top of the rectangular painting for shipping. What size and shape box can he use?

   _____

Name _____

Circles

Identify the parts of circle H.

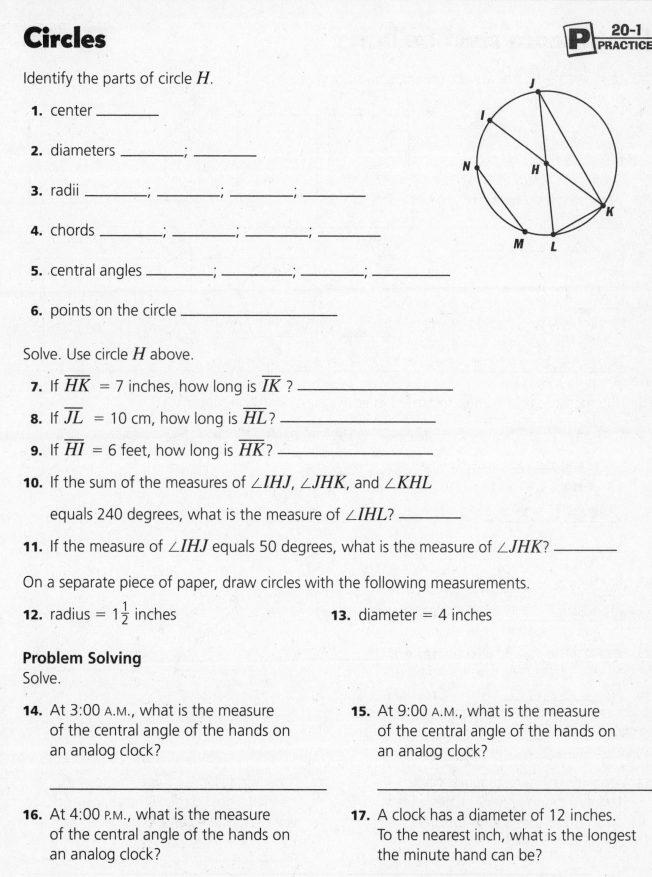

1. center _____

2. diameters _____; _____

3. radii _____; _____; _____; _____

4. chords _____; _____; _____; _____

5. central angles _____; _____; _____; _____

6. points on the circle _____

Solve. Use circle H above.

7. If $\overline{HK}$ = 7 inches, how long is $\overline{IK}$? _____

8. If $\overline{JL}$ = 10 cm, how long is $\overline{HL}$? _____

9. If $\overline{HI}$ = 6 feet, how long is $\overline{HK}$? _____

10. If the sum of the measures of $\angle IHJ$, $\angle JHK$, and $\angle KHL$

equals 240 degrees, what is the measure of $\angle IHL$? _____

11. If the measure of $\angle IHJ$ equals 50 degrees, what is the measure of $\angle JHK$? _____

On a separate piece of paper, draw circles with the following measurements.

12. radius = $1\frac{1}{2}$ inches

13. diameter = 4 inches

Problem Solving
Solve.

14. At 3:00 A.M., what is the measure of the central angle of the hands on an analog clock?

_____

15. At 9:00 A.M., what is the measure of the central angle of the hands on an analog clock?

_____

16. At 4:00 P.M., what is the measure of the central angle of the hands on an analog clock?

_____

17. A clock has a diameter of 12 inches. To the nearest inch, what is the longest the minute hand can be?

_____

Name _____

Congruence and Similarity

Tell whether the figures are congruent, similar, or neither.

1. _____

2. _____

3. _____

4. _____

5. _____

6. _____

Find the measure of the angle indicated in each pair of similar figures.

7.

100°
50°   30°
?

_____

8.

135°   45°
45°   135°
?

_____

9.

60°
120°
?

_____

Identify the corresponding side or angle.

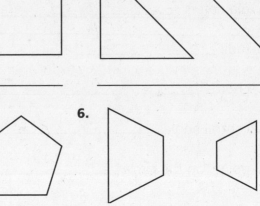

10. $\overline{GI}$ _____

11. $\angle P$ _____

12. $\overline{PQ}$ _____

13. $\angle R$ _____

14. $\overline{HI}$ _____

15. $\angle H$ _____

Problem Solving
Solve.

16. A poster is in a rectangular frame 16 inches long and 12 inches wide. What is the shape and size of a congruent frame?

_____

17. A painting is in a square frame with 18-inch sides. What is the shape and size of a similar frame?

_____

Name _____

Transformations

Graph a triangle with vertices (2, 4), (2, 6), and (5, 4).
Then transform the triangle to the new given vertices.
Write *translation, reflection,* or *rotation.*

1. (7, 4), (10, 6), and (10, 4)   **2.** (5, 7), (7, 7), and (5, 4)   **3.** (6, 1), (6, 3), and (9, 1)

_____   _____   _____

Write the combination of transformations used.

4.    **5.**   **6.**

_____   _____   _____

_____   _____   _____

Draw each figure after the transformation described.

7. rotation   **8.** translation   **9.** reflection

Problem Solving
Solve.

10. Which uppercase letters look like different uppercase letters when they are reflected over a horizontal line?

_____

11. You can transform a lowercase *v* in two different ways so that the *v* and its transformations form other lowercase letters. Describe the transformations and tell what letters you would form.

_____

_____

_____

Name _____

Problem Solving: Strategy
Find a Pattern

Find a pattern to solve each problem. State the pattern you followed.

1. Martina is designing chains. The diagram shows the number of rings she uses in each chain. If she continues the pattern, how many rings will be in the next chain?

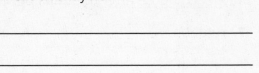

_____

2. Art A sculptor is using a pattern of different size cubes to create a sculpture with four sections. The first section has 1 cube, the second section has 16 cubes, and the third section has 81 cubes. How many cubes are in the fourth section?

_____

3. Mika is making a pattern of circles. The smallest circle has a diameter of 8 centimeters. The next circle has a diameter of 12 centimeters and the circle after that has a diameter of 16 centimeters. What is the diameter of the sixth circle?

_____

_____

4. The bottom layer of a pyramid has 216 blocks. The layer above the bottom has 125 blocks. The third layer from the bottom has 64 blocks. If the pattern continues, how many blocks will be in the next two layers?

_____

_____

Mixed Strategy Review
Solve. Use any strategy.

5. Number Sense In a set of bowls, the difference in the diameter between one bowl and the next size is 5 centimeters. The largest bowl has a diameter of 40 centimeters. If the smallest bowl has a diameter of 15 centimeters, how many bowls are in the set?

_____

Strategy: _____

6. Jessica buys a combination of 8 erasers and pencils. Pencils cost $0.18 each and erasers cost $0.35 each. The total cost of the pencils and the erasers is $1.78. How many of each did she buy?

_____

Strategy: _____

_____

_____

_____

7. Write a problem which you could find a pattern to solve. Share it with others.

Use with Grade 5, Chapter 20, Lesson 4, pages 482–483.

_____

Perimeter of Polygons

Find the perimeter of each figure.

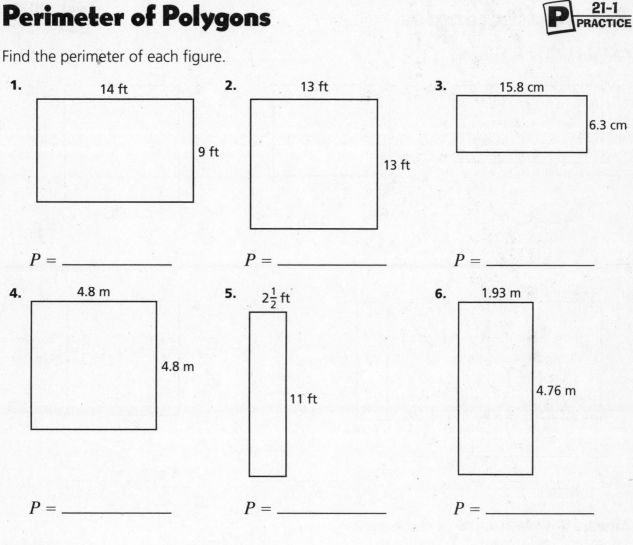

1. 14 ft 9 ft

P = _____

2. 13 ft 13 ft

P = _____

3. 15.8 cm 6.3 cm

P = _____

4. 4.8 m 4.8 m

P = _____

5. $2\frac{1}{2}$ ft 11 ft

P = _____

6. 1.93 m 4.76 m

P = _____

Algebra Find each missing measurement.

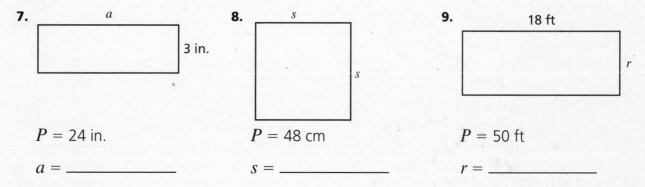

7. a 3 in.

P = 24 in.

a = _____

8. s s

P = 48 cm

s = _____

9. 18 ft r

P = 50 ft

r = _____

Problem Solving
Solve.

10. A square patio is 18 feet on each side. What is the perimeter of the patio?

_____

11. A rectangular garden is 9.7 meters long and 6 meters wide. How many meters of fencing are needed to enclose the garden?

_____

Use with Grade 5, Chapter 21, Lesson 1, pages 498–500.

Area of Rectangles

Find the area of each figure.

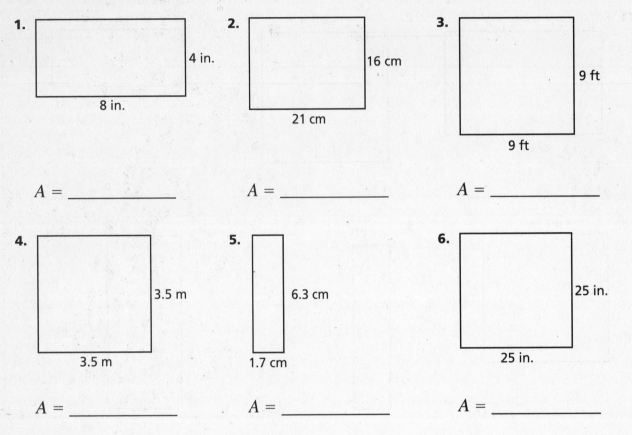

1. 4 in. 8 in.

    A = _____

2. 16 cm 21 cm

    A = _____

3. 9 ft 9 ft

    A = _____

4. 3.5 m 3.5 m

    A = _____

5. 6.3 cm 1.7 cm

    A = _____

6. 25 in. 25 in.

    A = _____

Algebra Find each missing measurement.

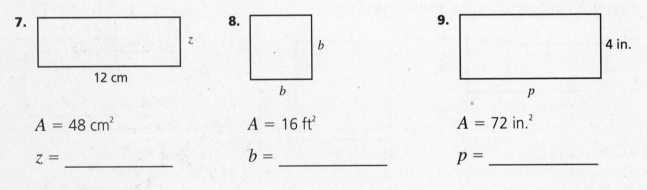

7. 12 cm z

    A = 48 cm²

    z = _____

8. b b

    A = 16 ft²

    b = _____

9. 4 in. p

    A = 72 in.²

    p = _____

Problem Solving
Solve.

10. A family room is 24 feet long and 18 feet wide. What is the area of the family room?

    _____

11. A square carpet is 3.6 meters on each side. What area will the carpet cover?

    _____

Change the Figure

Find the area and the perimeter of each figure. Then combine the figures to make a rectangle. Find the area and perimeter of the new rectangle.

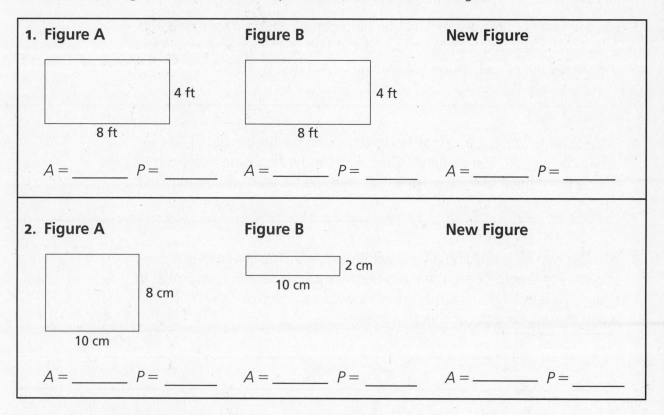

1. Figure A   **Figure B**   **New Figure**

4 ft   4 ft

8 ft   8 ft

A = _____ P = _____   A = _____ P = _____   A = _____ P = _____

2. Figure A   **Figure B**   **New Figure**

2 cm

10 cm

8 cm

10 cm

A = _____ P = _____   A = _____ P = _____   A = _____ P = _____

Find the area and perimeter of the figure that is formed when you combine the figure with a figure that has the same size and shape.

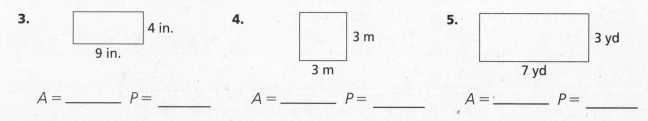

3.   **4.**   **5.**

4 in.   3 m   3 yd

9 in.   3 m   7 yd

A = _____ P = _____   A = _____ P = _____   A = _____ P = _____

Find the area and perimeter of the figure that is formed when you divide the figure into two congruent shapes.

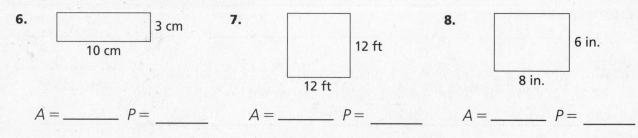

6.   **7.**   **8.**

3 cm   12 ft   6 in.

10 cm   12 ft   8 in.

A = _____ P = _____   A = _____ P = _____   A = _____ P = _____

Name _____

Problem Solving: Skill
Distinguish Between Perimeter and Area

Solve. State whether you need to find perimeter or area.

1. Hayden wants to make a rectangular herb garden that is 4 feet long and
   3 feet wide. She wants to plant lavender in half of the garden. How can
   she decide how much of the garden will be covered with lavender?

   _____

2. Daniel wants to plant a row of marigolds along the border of his
   vegetable garden. The garden is 6 feet long and 4 feet wide. How can he
   decide how much of the garden will need to be covered with marigolds?

   _____

3. Ms. Carmichael is building a deck with two levels. The lower level is a
   square. The length of each side is 5 feet. The upper level is rectangular in
   shape, 12 feet long and 8 feet wide. How can she decide how much
   wood she will need to construct each level?

   _____

   _____

   _____

4. Ms. Carmichael wants to put railing around the sides of the lower level.
   How can she decide how much railing she will need?

   _____

5. Jamison has 70 square feet of plywood to make a floor for a two-room
   clubhouse he is building. The floor of one room is 8 feet long and 6 feet
   wide. The floor of the other room is 5 feet long and 4 feet wide. How can
   he decide if he has enough plywood?

   _____

   _____

6. Amy wants to make a frame for a painting that is 24 inches long and
   18 inches wide. She found a wood molding she would like to use. How
   can she decide how much molding she needs to make the frame?

   _____

Use with Grade 5, Chapter 21, Lesson 4, pages 508–509.

Name _____

Explore Area of Parallelograms

Find the area of each figure.

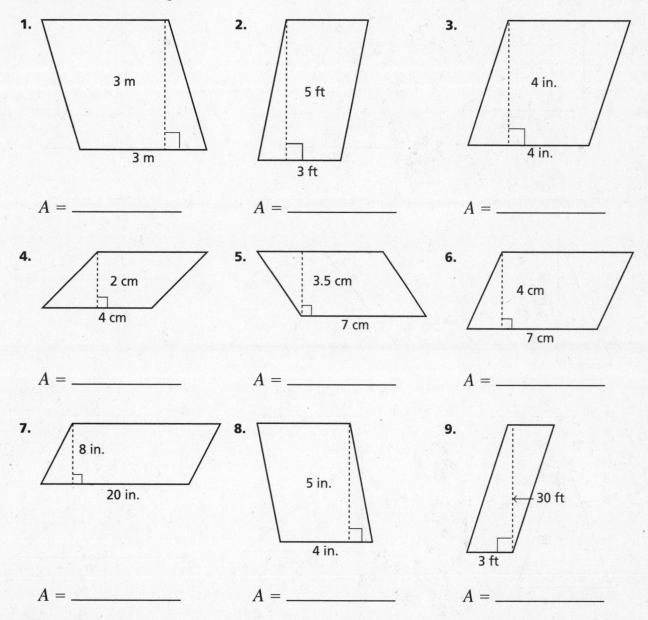

1. 3 m / 3 m

A = _____

2. 5 ft / 3 ft

A = _____

3. 4 in. / 4 in.

A = _____

4. 2 cm / 4 cm

A = _____

5. 3.5 cm / 7 cm

A = _____

6. 4 cm / 7 cm

A = _____

7. 8 in. / 20 in.

A = _____

8. 5 in. / 4 in.

A = _____

9. 30 ft / 3 ft

A = _____

Problem Solving
Solve.

10. A garden in the shape of a parallelogram has a base of 14 meters and a height of 3 meters. What is the area of the garden?

_____

11. Another garden in the shape of a parallelogram covers 76 square feet. Its height is 4 feet. What is its base?

_____

Explore Area of Triangles

Find the area of each triangle.

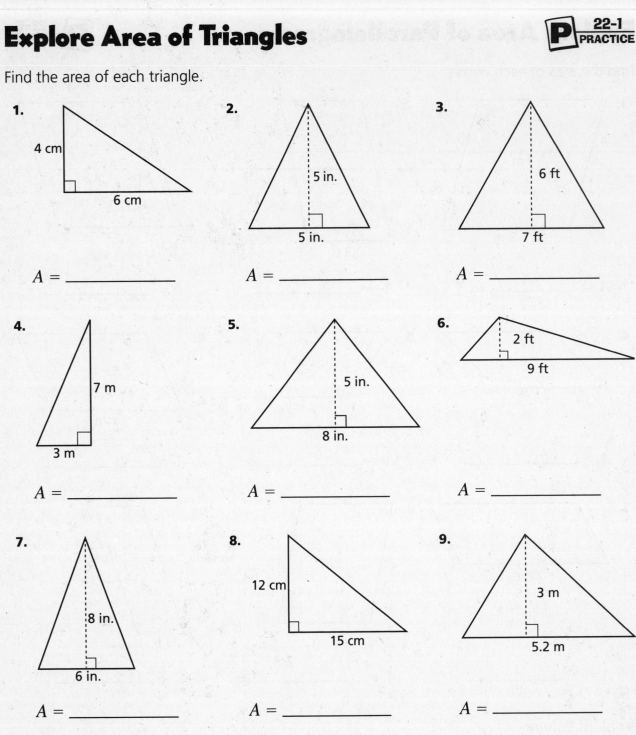

1. 4 cm, 6 cm

A = _____

2. 5 in., 5 in.

A = _____

3. 6 ft, 7 ft

A = _____

4. 7 m, 3 m

A = _____

5. 5 in., 8 in.

A = _____

6. 2 ft, 9 ft

A = _____

7. 8 in., 6 in.

A = _____

8. 12 cm, 15 cm

A = _____

9. 3 m, 5.2 m

A = _____

Problem Solving
Solve.

10. The triangular sail on a boat has a base of 8 feet and a height of 12 feet. What is the area of the sail?

_____

11. A triangular flag has a base of 18 centimeters and a height of 30 centimeters. What is the area of the flag?

_____

Use with Grade 5, Chapter 22, Lesson 1, pages 516–517.

Name _____

Explore Area of Trapezoids

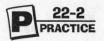
Find the area of each trapezoid

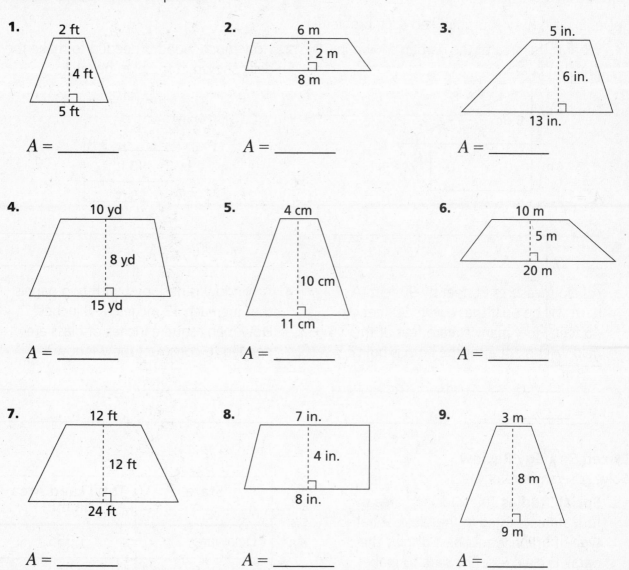

1. 2 ft
4 ft
5 ft

A = _____

2. 6 m
2 m
8 m

A = _____

3. 5 in.
6 in.
13 in.

A = _____

4. 10 yd
8 yd
15 yd

A = _____

5. 4 cm
10 cm
11 cm

A = _____

6. 10 m
5 m
20 m

A = _____

7. 12 ft
12 ft
24 ft

A = _____

8. 7 in.
4 in.
8 in.

A = _____

9. 3 m
8 m
9 m

A = _____

Problem Solving
Solve.

10. The shape of a garden is a trapezoid with bases of 8 feet and 10 feet. The height of the trapezoid is 6 feet. What area does the trapezoid cover?

_____

11. A window is a trapezoid with bases of 4 feet and 6 feet. The height of the trapezoid is 5 feet. How much glass would be needed to replace the window?

_____

Name _____

Problem Solving: Strategy
Solve a Simpler Problem

Solve. Explain how you simplified each problem.

1. What is the area of the garden shown in the plan below?

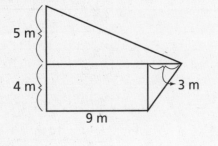

5 m 4 m 3 m 9 m

_____

_____

2. How much wood is needed to make the deck shown in the plan below?

9 ft
5 ft
3 ft
3 ft
2 ft 9 ft

_____

_____

3. A field measures 80 feet by 90 feet. A barn will be built that covers 30 feet by 45 feet. How many square feet of the field will be left after the barn is built?

_____

_____

4. A window is designed with two panels that are each 12 inches by 8 inches. How many square inches of glass are needed to construct the window?

_____

_____

Mixed Strategy Review
Solve. Use any strategy.

5. Social Studies The total land area of four states is listed in the table. What type of graph would best display the data? Explain. Use the data to make the graph.

_____

Strategy: _____

6. Health Leo increases the number of push-ups he does each week by 8. The first week he did 10 push-ups. He is now doing 42 push-ups. How many weeks has he been doing push-ups?

_____

Strategy: _____

State	Total Land Area (in mi²)
Delaware	1,955
New Hampshire	8,969
Rhode Island	1,045
Connecticut	4,845

7. Write a problem that you could solve with a simpler problem. Share it with others.

_____

_____

Perimeter and Area of Irregular Figures

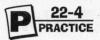

Find the perimeter of each figure. Each square = 1 in.²

1.
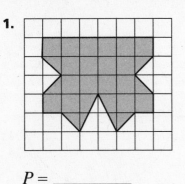

P = _____

2.
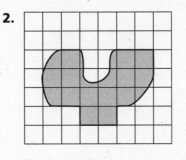

P = _____

3.
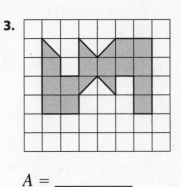

A = _____

Count or estimate to find the area. Each square = 1 cm².

4.
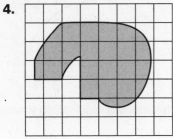

A = _____

5.
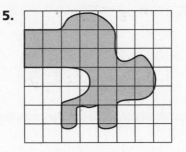

A = _____

6.
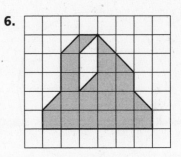

A = _____

7.
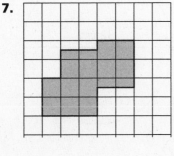

A = _____

8.
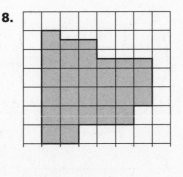

A = _____

9.
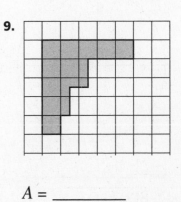

A = _____

Explore Circumference of Circles

Find the approximate circumference of each circle. Use $\pi \approx 3.14$.
Round to the nearest tenth, if necessary.

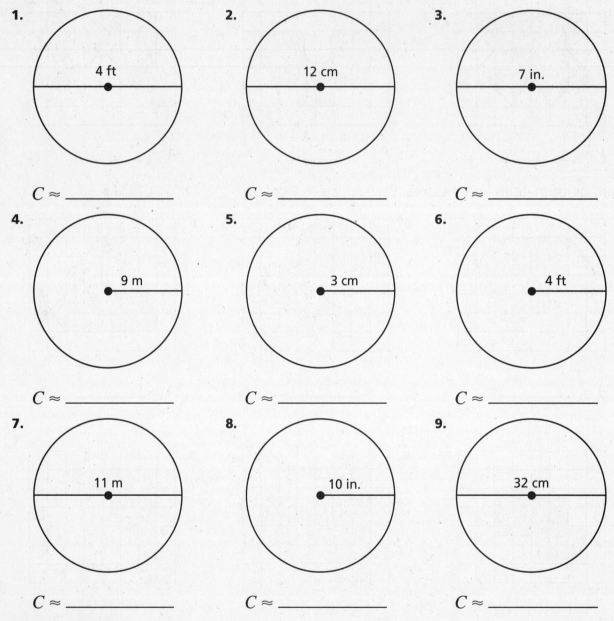

1.

4 ft

$C \approx$ _____

2.

12 cm

$C \approx$ _____

3.

7 in.

$C \approx$ _____

4.

9 m

$C \approx$ _____

5.

3 cm

$C \approx$ _____

6.

4 ft

$C \approx$ _____

7.

11 m

$C \approx$ _____

8.

10 in.

$C \approx$ _____

9.

32 cm

$C \approx$ _____

Problem Solving
Solve.

10. A swimming pool has a diameter of 22 feet. To the nearest tenth of a foot, what is the circumference of the pool?

_____

11. A fountain is directly in the center of a circular pool. It is 8 meters from the wall surrounding the pool. To the nearest tenth of a meter, what is the length of the wall of the pool?

_____

Use with Grade 5, Chapter 22, Lesson 5, pages 526–527.

3-Dimensional Figures and Nets

Write the number of faces, edges, and vertices for each figure.

1. Faces: _____   Edges: _____   Vertices: _____

2. Faces: _____   Edges: _____   Vertices: _____

3. Faces: _____   Edges: _____   Vertices: _____

4. Faces: _____   Edges: _____   Vertices: _____

What 3-dimensional figure does each net make when cut and folded?

5.

6.

_____   _____

Problem Solving

Solve. Use data from the art for problems 7 and 8.

7. What shape was used for the bottom part of the building?

_____

8. What shape was used for the top part of the building?

_____

3-Dimensional Figures from Different Views

Draw the top view, front view, and a side view of the shape.

	Top	Front	Side

1.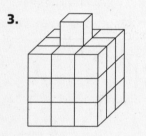

2.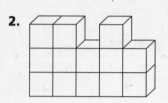

3.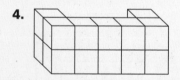

4.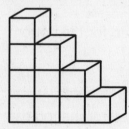

Problem Solving

Solve.

5. This staircase is made from cubes. Draw the top view, front view, and side view of the staircase.

Explore Surface Area of Rectangular Prisms

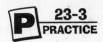

Find the surface area of each rectangular prism.

1.

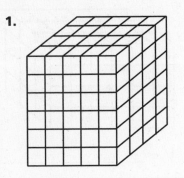

_____

2.

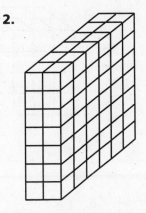

_____

3.

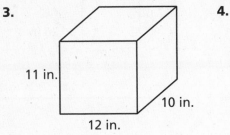

11 in.

12 in.

10 in.

_____

4.

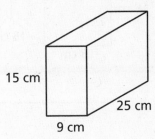

15 cm

9 cm

25 cm

_____

5.

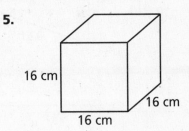

16 cm

16 cm

16 cm

_____

6.

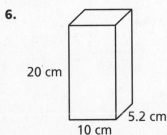

20 cm

10 cm

5.2 cm

_____

7.

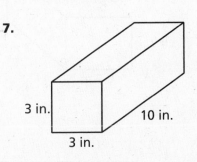

3 in.

3 in.

10 in.

_____

8.

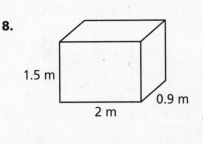

1.5 m

2 m

0.9 m

_____

Problem Solving
Solve.

9. What is the surface area of a cardboard shipping box that is 26 inches long, 26 inches wide, and 18 inches high?

_____

10. What is the surface area of a 9-centimeter cube?

_____

Volume of Rectangular Prisms

Find the volume of each rectangular prism.
Round to the nearest tenth, if necessary.

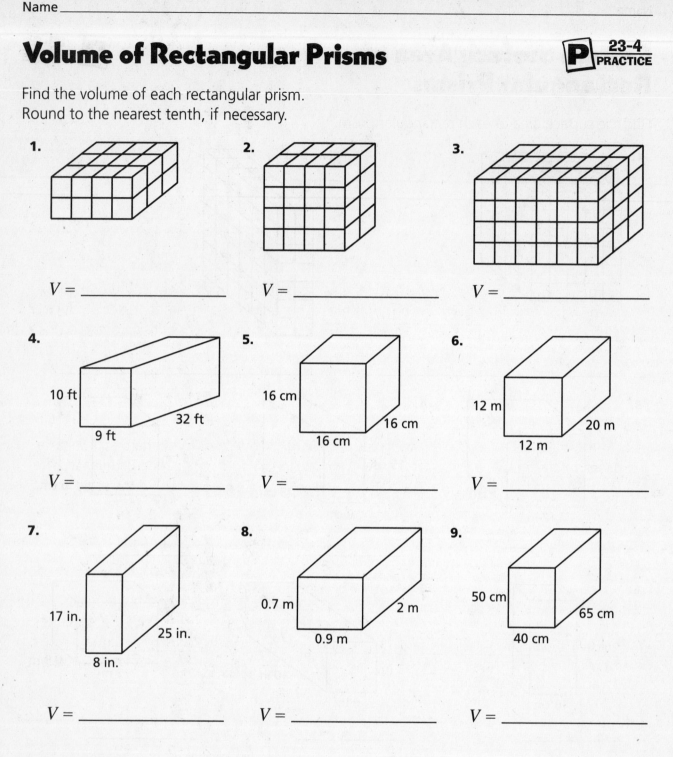

1.

V = _____

2.

V = _____

3.

V = _____

4.

10 ft
32 ft
9 ft

V = _____

5.

16 cm
16 cm
16 cm

V = _____

6.

12 m
20 m
12 m

V = _____

7.

17 in.
25 in.
8 in.

V = _____

8.

0.7 m
2 m
0.9 m

V = _____

9.

50 cm
65 cm
40 cm

V = _____

Problem Solving
Solve.

10. The dimensions of a gift box for jewelry are 6 inches by 3 inches by 2 inches. What is the volume of the gift box?

_____

11. The dimensions of a shoe box are 13 inches by 9 inches by 4 inches. What is the volume of the shoe box?

_____

Name _____

Problem Solving: Skill
Follow Directions

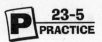

Follow the directions to solve the problem.

1. The top of a building is a prism that is 12 m long, 15 m wide, and 20 m tall. The base is a cube that is 25 m on each side. Draw a model of the building. Let 1 cm in the drawing equal 1 m of the building. What is the volume of the building?

_____

2. A wooden box has two sections. The top section is a rectangular prism that is 5 in. long, 12 in. wide, and 3 in. high. The bottom section is a rectangular prism that is 5 in. long, 12 in. wide, and 8 in. high. Draw a model of the box. Suppose you want to paint the outside of the box. What area will you need to cover?

_____

3. Jean makes a sculpture out of clear plastic. She fills the sculpture with marbles. The base of the sculpture is a rectangular prism that is 20 cm tall, 10 cm long, and 8 cm wide. The top of the sculpture is a cube that is 5 cm on each side. Draw a model of the sculpture. What is the volume of the sculpture?

_____

4. A plot of land contains buildings that are congruent rectangular prisms that are each 30 m tall, 15 m wide, and 15 m long. Three sides of each building are flush with the edges of the plot of land. The fourth side of one building faces the fourth side of the other building. These two sides are 10 m apart. Draw a model of the two buildings and the plot of land. What is the area of the plot of land?

_____

5. Four cubes are placed in a row. The cubes form a rectangular prism. Each cube is 8 in. on a side. Draw a diagram of the figure that is made from the cubes. What is the surface area of this figure?

_____

Line Symmetry

Tell which figures are symmetric about a line. Draw lines of symmetry.

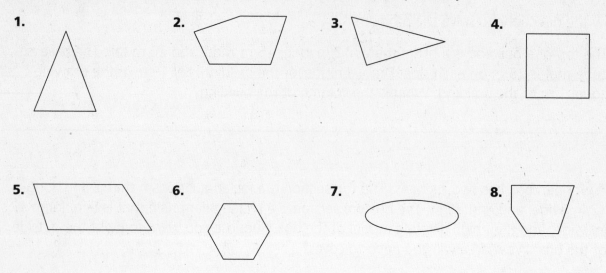

1.   2.   3.   4.

5.   6.   7.   8.

Complete each figure so that the line is a line of symmetry.

9.   10.   11.

Problem Solving
Solve.

12. Print your name in uppercase letters.
Then print your name in lowercase letters.
Show which letters are symmetric.

_____

_____

13. Draw a polygon with more than four
sides that has at least one line of
symmetry. Name the polygon you draw.

_____

_____

Use with Grade 5, Chapter 24, Lesson 1, pages 560–562.

Rotational Symmetry

Tell whether the figure has rotational symmetry. If it does, find the smallest fraction of a full turn needed for the figure to look the same.

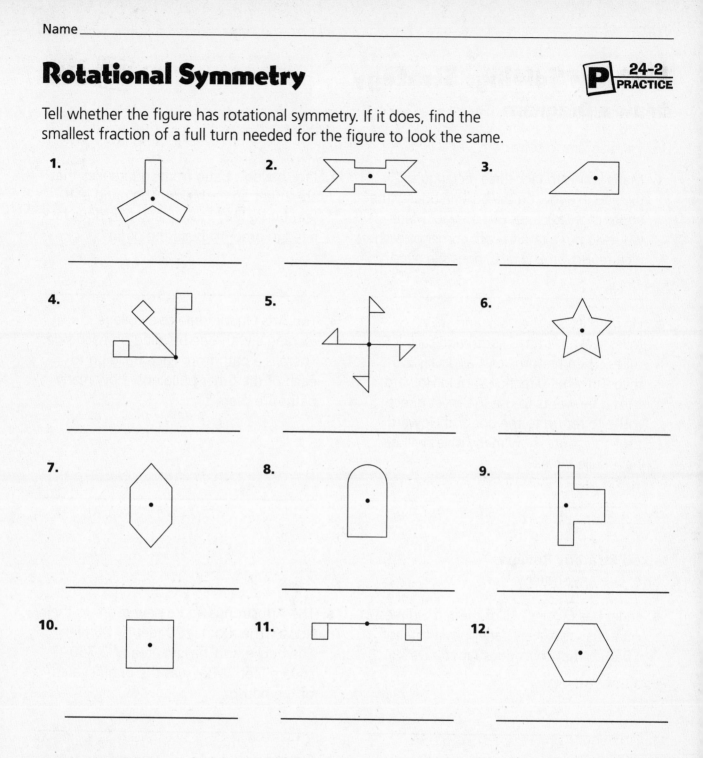

1.

_____

2.

_____

3.

_____

4.

_____

5.

_____

6.

_____

7.

_____

8.

_____

9.

_____

10.

_____

11.

_____

12.

_____

Problem Solving
Solve.

13. Draw a figure that has rotational symmetry. Name the smallest fraction of a full turn needed for the figure to look the same.

_____

14. What is the first capital letter of the alphabet that has rotational symmetry? Name the smallest fraction of a full turn needed for the figure to look the same.

_____

Name _____

Problem Solving: Strategy
Draw a Diagram

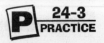

Draw a diagram to solve.

1. Maria wants to tack three rectangular pictures in a row on the bulletin board. The edges of the pictures can overlap. Maria wants to put a tack in each corner of each picture. How many tacks does she need?

_____

2. Jack builds a patio from square tiles that are 2 feet on each side. The patio is 10 feet by 16 feet. How many tiles does Jack need in order to build the patio?

_____

3. Howard leaves the dock and sails 2.5 miles west. He turns south and sails 3.5 miles. Then he turns east and sails 2.5 miles. In what direction should Howard turn if he wants to use the most direct route to return to the dock? If Howard uses this route, how many miles will he have sailed in all?

_____

4. The Arts Quadrangle of a college is a rectangle with one building on each side. There is a path from each building to each of the other buildings. How many paths are there?

_____

Mixed Strategy Review
Solve. Use any strategy.

5. Janet has 6 coins. All of them are dimes, quarters, or nickels. Janet has a total of $0.65. What coins does Janet have?

_____

Strategy: _____

6. The Armstrongs add a new room to their house. The room is 20 feet by 25 feet. The house now has an area of 2,750 square feet. What was the original area of the house?

_____

Strategy: _____

7. The School Fund raises $2,300. After paying for new sports equipment, the fund has $1,170. How much did the School Fund spend on sports equipment?

_____

Strategy: _____

8. **Write a problem** that you would draw a diagram to solve. Share it with others.

_____

_____

Use with Grade 5, Chapter 24, Lesson 3, pages 566–567.

Name _____

Explore Tessellations

Tell whether each shape tessellates. Record your work.

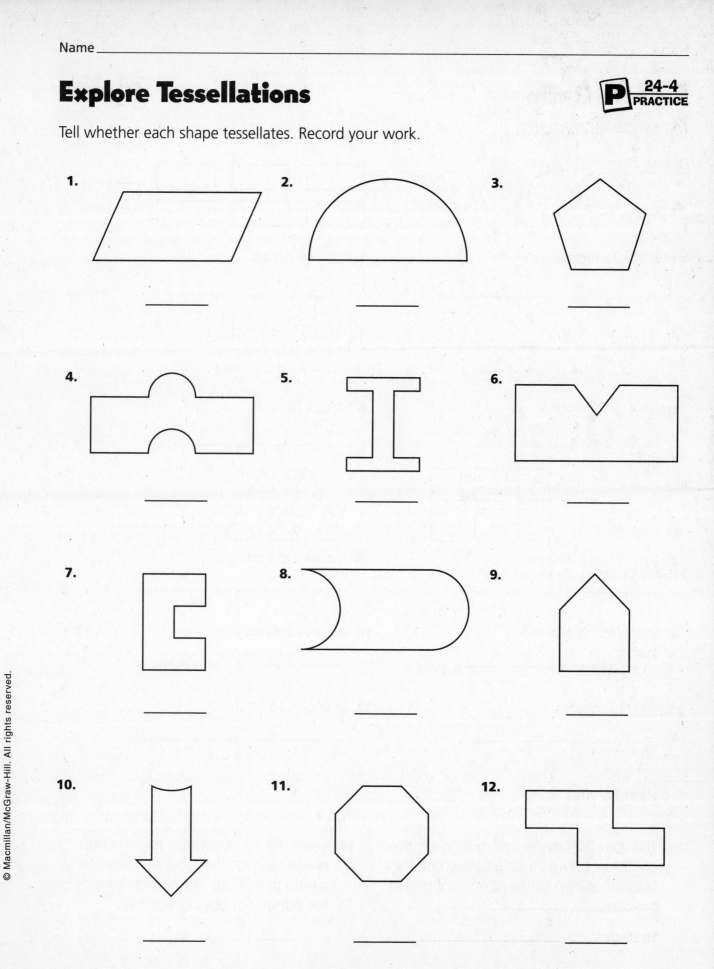

1. _____

2. _____

3. _____

4. _____

5. _____

6. _____

7. _____

8. _____

9. _____

10. _____

11. _____

12. _____

Use with Grade 5, Chapter 24, Lesson 4, pages 568–569.

Explore Ratio

Write each ratio in three ways.

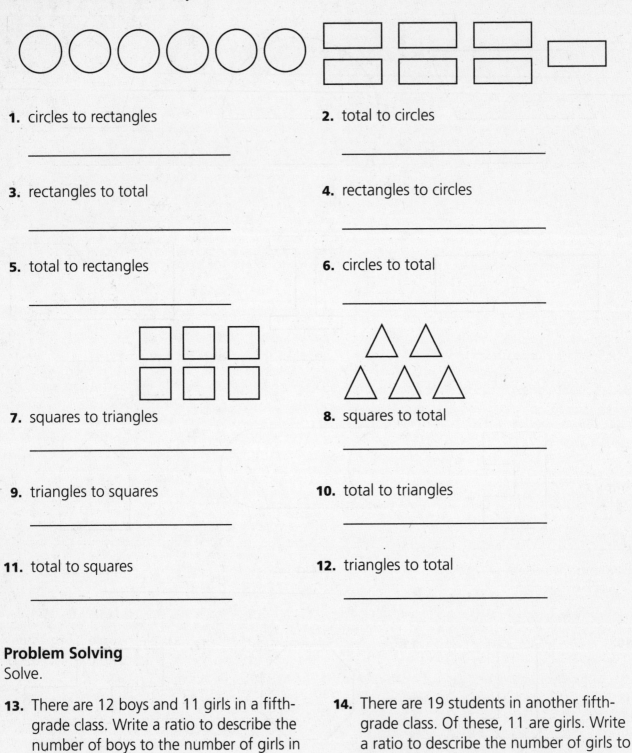

1. circles to rectangles

_____

2. total to circles

_____

3. rectangles to total

_____

4. rectangles to circles

_____

5. total to rectangles

_____

6. circles to total

_____

7. squares to triangles

_____

8. squares to total

_____

9. triangles to squares

_____

10. total to triangles

_____

11. total to squares

_____

12. triangles to total

_____

Problem Solving

Solve.

13. There are 12 boys and 11 girls in a fifth-grade class. Write a ratio to describe the number of boys to the number of girls in the class.

_____

14. There are 19 students in another fifth-grade class. Of these, 11 are girls. Write a ratio to describe the number of girls to the number of boys in this class.

_____

Use with Grade 5, Chapter 25, Lesson 1, pages 584–585.

Algebra: Equivalent Ratios

Complete each ratio table.

1.

1		3		5
4	8		16	

2.

5			20	25
7	14	21		

3.

9	18		36	
4		12		20

4.

10		30		50
3	6		12	

Tell whether the ratios are equivalent. Write *yes* or *no*.

5. $\frac{2}{5}$, $\frac{8}{20}$ _____

6. $\frac{6}{7}$, $\frac{30}{42}$ _____

7. $\frac{20}{12}$, $\frac{4}{3}$ _____

8. 15:9, 5:3 _____

9. 4:10, 30:12 _____

10. 5:8; 25:40 _____

Name four ratios equivalent to each given ratio.

11. $\frac{1}{3}$ _____

12. $\frac{4}{5}$ _____

13. $\frac{7}{2}$ _____

14. $\frac{9}{8}$ _____

15. $\frac{80}{20}$ _____

16. $\frac{25}{75}$ _____

Find the missing number.

17. 4:7 = c:35

$c =$ _____

18. 3:8 = 27:s

$s =$ _____

19. 32:12 = h:3

$h =$ _____

20. 7 to 2 = 42 to d

$d =$ _____

21. 65 to 25 = k:5

$k =$ _____

22. 5 to 11 = 40 to m

$m =$ _____

Problem Solving

Solve.

23. One store has 3 managers and 12 salespeople. Another store has 4 managers and 15 salespeople. Do both stores have equivalent ratios of managers to salespeople? Explain.

_____

_____

_____

24. A store uses the ratio 1 to 5 as a guide for managers to salespeople. Suppose the store has 30 salespeople. How many managers should it have?

_____

Name _____

Rates

Complete.

1. 120 mi in 3 h = _____ mi in 9 h

2. 27 pages in 2 d = _____ pages in 10 d

3. 7 problems in 10 min = _____ problems in 60 min

4. 10 oz for 3 people = _____ oz for 12 people

5. 3 books in 2 wk = 24 books in _____ wk

6. 16 people in 1 van = _____ people in 3 vans

7. $15 for 2 tickets = _____ for 20 tickets

8. 5 for $1.99 = 20 for _____

9. $45 for 3 CDs = _____ for 6 CDs

10. 50 pennies in 1 roll = 200 pennies in _____ rolls

11. $25 in 2 h = _____ in 8 h

12. 6 for $15 = _____ for $150

13. 32 students in 4 groups = 128 students in _____ groups

14. 250 pieces of paper in 2 packs = 1,000 pieces of paper in _____ packs

Find each unit rate.

15. 35 people in 7 cars = _____ people per 1 car

16. 175 words in 5 min = _____ words in 1 min

17. $4.96 for 16 oz = _____ per 1 oz

18. 210 mi in 4 h = _____ mi per 1 h

19. 192 mi on 8 gal = _____ mi per 1 gal

20. 15 in. of rain in 30 d = _____ in. per 1 d

21. $40.50 for 9 tickets = _____ for 1 ticket

22. $49.50 for 6 h = _____ for 1 h

23. $75 for 3 shirts = _____ per 1 shirt

24. 96 pages in 6 min = _____ pages per 1 min

25. 248 mi on 8 gal = _____ mi on 1 gal

26. 30 passengers in 5 vans = _____ passengers in 1 van

27. $17 for 4 lb = _____ for 1 lb

28. $18.75 for 5 lb = _____ for 1 lb

Problem Solving
Solve.

29. A $3.36 box of cereal contains 14 servings. What is the cost per serving?

30. Enough bread for 10 sandwiches costs $1.89. How much will enough bread for 80 sandwiches cost?

_____

_____

Better Buy

Find the unit price. Round to the nearest cent.

1. 3 liters for $7.29

_____

2. 5 pens for $2.19

_____

3. 16 ounces for $20

_____

4. 7 hours for $85

_____

5. 4 tons for $560

_____

6. 6 tubes for $18.50

_____

Explain which is the better buy.

7. 20 disks for $5

50 disks for $12

_____

8. 10 pens for $4.50

3 pens for $1.47

_____

9. 5 pounds for $2.99

12 pounds for $7.29

_____

10. 12 granola bars for $5.75

8 granola bars for $3.75

_____

11. 6 eggs for $0.84

1 dozen eggs for $1.80

_____

12. 2 rolls of film for $5.39

3 rolls of film for $8

_____

Problem Solving
Solve.

13. Mika's Messenger Service buys 4 bicycles for $500. Zip Messengers, Inc., buys 6 bicycles of the same type for $725. Which messenger service gets the better buy?

_____

14. A box of 25 pencils costs $4.75. A box of 40 pencils has a unit price that is $0.02 less. What is the price of the box of 40 pencils?

_____

Name _____

Problem Solving: Skill
Check the Reasonableness of an Answer

Read the details in each problem. Check whether each solution is reasonable. Explain.

1. Cynthia and Marcy are starting a pet-sitting business. While pet owners are out of town, Cynthia and Marcy will take care of each pet for $5.00 a day. This week, they will watch 3 pets for 4 days. They decide that together they will make $30 for the week. Is their estimate reasonable?

   _____

   _____

2. Tasha earns extra money by babysitting on weekends. For each child she watches, she charges $4.50 per evening. On Friday evening she watches 4 children and on Saturday evening she watches 5 children. She thinks she will earn about $40. Is her estimate reasonable?

   _____

   _____

3. Chandler charges $10.00 to mow a lawn. This week, he is scheduled to mow 2 lawns each day from Monday through Saturday. He decides that he will make $70. Is his estimate reasonable?

   _____

   _____

4. Ricky hangs sale advertisements in a grocery store window. He earns $2.25 for each advertisement that he hangs up. On Monday, he hangs up 6 advertisements. On Friday, he hangs up 3 advertisements with the weekend specials. On Saturday, he hangs up 2 more advertisements. He thinks he will earn $16. Is his estimate reasonable?

   _____

   _____

5. To raise money for her school, Amy sells gift-wrapping paper. She thinks she can sell 8 rolls per day. She can only sell the gift-wrapping paper from Thursday through Saturday. The paper costs $3 per roll. She decides she will raise about $70. Is her estimate reasonable?

   _____

   _____

Use with Grade 5, Chapter 25, Lesson 5, pages 596–597.

Algebra: Scale Drawings

Use data from the floor plan. Find each actual size.

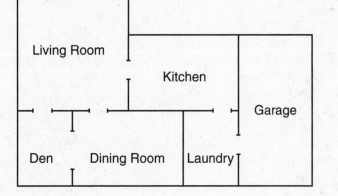

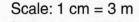
Scale: 1 cm = 3 m

1. length of garage _____

2. width of garage _____

3. width of doorway from den to dining room _____

4. perimeter of kitchen _____

5. width of the house _____

6. length of living room and den together _____

Find the scale.

7. An actual family room that is 8 meters long is 4 centimeters on a drawing.

_____

8. An actual closet that is 6 feet long is 2 inches on a drawing.

_____

9. An actual bedroom that is 15 feet long is 3 inches on a drawing.

_____

10. An actual house that is 16 meters wide is 4 centimeters on a drawing.

_____

Problem Solving
Solve.

11. A map has a scale of 1 inch = 4 miles. The map distance from Jacob's house to his school is $2\frac{1}{2}$ inches. How far does Jacob actually live from his school?

_____

12. Shannon lives 15 kilometers from her school. She plans to draw a map using the scale 1 centimeter = 5 kilometers to show this. How far will her house be from the school on her map?

_____

Name _____

Explore Probability

Use the spinner for problems 1–6.

1. What are the possible outcomes?

   _____

2. Which outcome is likely?

   _____

3. Which outcome is unlikely?

   _____

4. If you spin the spinner 30 times, what outcome do you think you will get most often?

   _____

5. If you spin the spinner 30 times, what outcome do you think you will get least often?

   _____

6. What if you make 3 of the striped sections speckled? Which outcome will be likely? Unlikely?

   _____

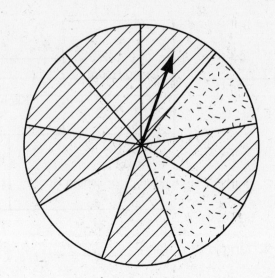

Use the bag of cubes for problems 7–12.

7. How many possible outcomes are there if you pick a cube out of the bag with your eyes closed? What are they?

   _____

8. Which outcome is likely?

   _____

9. Which outcome is unlikely?

   _____

10. If you pick a cube out of the bag 40 times, what cube do you think you will get most often?

    _____

11. If you pick a cube out of the bag 40 times, what cube do you think you will get least often?

    _____

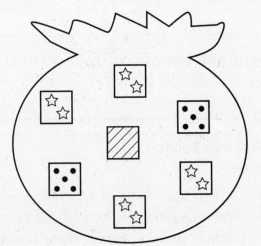

12. What cubes could you add to the bag to make picking a striped cube very likely?

    _____

Probability

If you spin this spinner, what is the probability of each event?

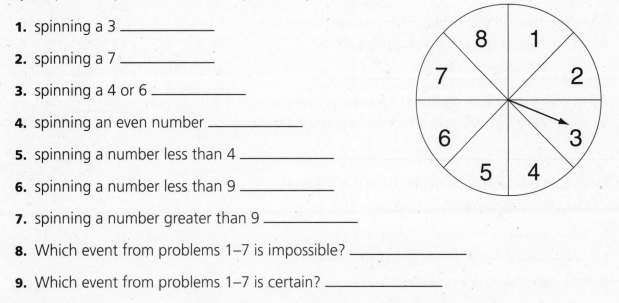

1. spinning a 3 _____

2. spinning a 7 _____

3. spinning a 4 or 6 _____

4. spinning an even number _____

5. spinning a number less than 4 _____

6. spinning a number less than 9 _____

7. spinning a number greater than 9 _____

8. Which event from problems 1–7 is impossible? _____

9. Which event from problems 1–7 is certain? _____

If you pick a card, what is the probability of each event? Write *more likely than*, *less likely than*, or *equally likely as* to complete each sentence.

10. Picking a circle is _____ picking a triangle.

11. Picking a square is _____ picking a triangle.

12. Picking a square is _____ picking a circle.

Problem Solving
Solve.

13. You and three friends are trying to decide what video to rent. You each write a different movie name on a card. If you pick a card at random, what is the probability that the movie name you wrote will be chosen?

_____

14. Two girls and three boys want to borrow the same book from the school library. Each writes his or her name on a card. If the librarian picks a card at random, what is the probability that a girl will be chosen to borrow the book?

_____

Name _____

Explore Making Predictions

Use the spinner at the right for problems 1–3.

1. If you spin the spinner 50 times, how many times do you predict you will spin an even number?

_____

2. If you spin the spinner 60 times, how many times do you predict you will spin a number less than 4?

_____

3. If you spin the spinner 80 times, how many times do you predict you will spin a number that is not 5?

_____

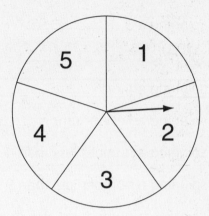

Use the spinner at right for problems 4–7.

4. If you spin the spinner 120 times, how many times do you predict you will spin a circle?

_____

5. If you spin the spinner 60 times, how many times do you predict you will spin a square?

_____

6. If you spin the spinner 150 times, how many times do you predict you will spin a circle or a hexagon?

_____

7. If you spin the spinner 90 times, how many times do you predict you will spin a triangle?

_____

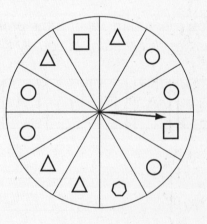

Name _____

Problem Solving: Strategy
Do an Experiment

Do an experiment to solve. Record the results in a frequency table.

1. **Language Arts** Which consonant is used most often in writing: *N, R, S,* or *T*? First, make a prediction. Then choose 20 lines of text from a book and count the number of each consonant to solve.

_____

2. **Language Arts** Which vowel (*A, E, I, O,* or *U*) is used most often to start a sentence? Predict which vowel you think is used the most. Then choose 50 sentences from a book and record the number of times each vowel starts a sentence to solve.

_____

3. What is the probability that a dropped dollar bill will land face down? First, make a prediction, and then record the results of dropping a dollar bill 50 times.

_____

4. **Literature** How often does the title of a novel include the word *the*? First, make a prediction. Then gather the titles of 20 novels and record the number of times each title includes the word *the*.

_____

Mixed Strategy Review
Solve. Use any strategy.

5. A music store sells new CDs for $12.75 and used CDs for $5.25 each. Blake buys 2 new CDs. He spends a total of $41.25. How many used CDs did he buy?

_____

Strategy: _____

6. Cindy has 500 coupons to hand out. She hands out about 125 coupons per hour. If she starts at 9 A.M., what time can she expect to be finished?

_____

Strategy: _____

7. **Literature** *The Boxcar Children,* a book by Gertrude Chandler Warner, has 153 pages. The book is being reprinted with a smaller type size. The new book will have $\frac{2}{3}$ the number of pages. How many pages will the new book have?

_____

Strategy: _____

8. **Write a problem** that you could do an experiment to solve. Share it with others.

_____

_____

_____

Explore Compound Events

Use the spinners at right for problems 1–3.

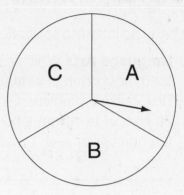

1. How many outcomes are possible if you spin both
   spinners? In the space below, draw a tree diagram
   to show them.

_____ possible outcomes

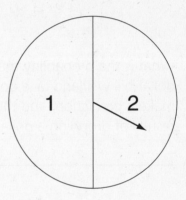

2. What is the probability of spinning C on the first
   spinner?

   _____

3. What is the probability of spinning 2 on the second
   spinner?

   _____

4. What is the probability of spinning C and 2 on one spin
   of each spinner?

   _____

Use the spinners at right for problems 5–7.

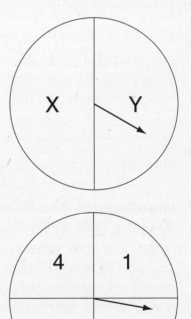

5. How many outcomes are possible if you spin both
   spinners?

   _____

6. What is the probability of spinning X and an even
   number on one spin of each spinner?

   _____

7. How many times do you predict you would spin X and
   an even number if you spin each spinner 40 times?

   _____

Use with Grade 5, Chapter 26, Lesson 5, pages 614–615.

Compound Events

Find the number of possible outcomes.

1. Spinning each spinner once

   _____ possible outcomes

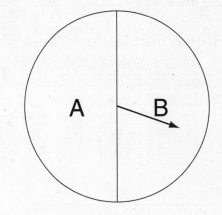

2. Buying a souvenir hat and T-shirt when hats come in 4 styles and T-shirts come in 2 styles

   _____ possible outcomes

3. Spinning a spinner with 8 sections labeled A, B, C, D, E, F, G, and H, and rolling a number cube with faces labeled 1 through 6

   _____ possible outcomes

4. Spinning a spinner with 10 sections labeled A, B, C, D, E, F, G, H, I, and J, and tossing a penny

   _____ possible outcomes

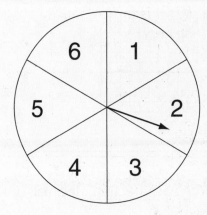

Solve.

5. What is the probability of spinning an A and a 3 with the spinners from problem 1?

   _____

6. What is the probability of spinning a C and tossing a heads with the spinner and the penny from problem 4?

   _____

7. What is the probability of spinning a vowel and tossing a heads with the spinner and the penny from problem 4?

   _____

Name _____

Explore the Meaning of Percent

Write a fraction, a decimal, a ratio, and a percent to show the shaded part of each grid. For each fraction, use simplest form.

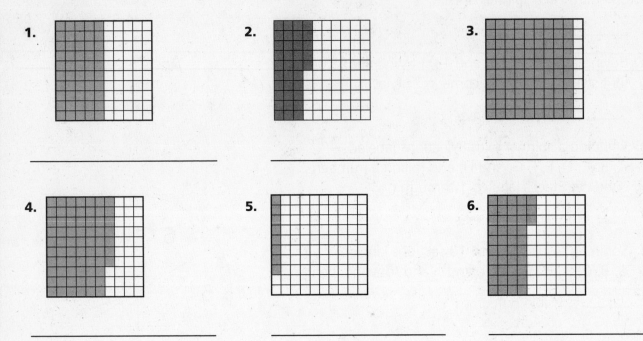

1. _____

2. _____

3. _____

4. _____

5. _____

6. _____

Write each fraction, decimal, or ratio as a percent.

7. $\frac{1}{4}$ _____

8. $\frac{17}{100}$ _____

9. $\frac{3}{10}$ _____

10. $\frac{1}{2}$ _____

11. 0.40 _____

12. 0.35 _____

13. 0.12 _____

14. 0.46 _____

15. 70:100 _____

16. 65:100 _____

17. 6:100 _____

18. 23:100 _____

19. $\frac{1}{10}$ _____

20. 0.19 _____

21. 99:100 _____

22. $\frac{9}{100}$ _____

23. 0.08 _____

24. $\frac{7}{10}$ _____

25. 0.77 _____

26. 10:100 _____

Problem Solving
Solve.

27. Three fourths of the shirts a store stocks are extra large. What percent of the shirts are extra large?

_____

28. Of the 100 shirts a store sold on Saturday, 82 had the logo of a sports team on them. What percent of the shirts had a logo?

_____

Use with Grade 5, Chapter 27, Lesson 1, pages 632–633.

Name _____

Percents, Fractions, and Decimals

Write each percent as a decimal and as a fraction in simplest form.

1. 34% _____   **2.** 70% _____   **3.** 48% _____

4. 25% _____   **5.** 7% _____   **6.** 45% _____

7. 12% _____   **8.** 54% _____   **9.** 91% _____

10. 95% _____   **11.** 32% _____   **12.** 82% _____

13. 57% _____   **14.** 24% _____   **15.** 30% _____

16. 18% _____   **17.** 72% _____   **18.** 88% _____

19. 60% _____   **20.** 22% _____   **21.** 96% _____

22. 9% _____   **23.** 35% _____   **24.** 61% _____

Write each fraction or decimal as a percent.

25. $\frac{3}{4}$ _____   **26.** $\frac{2}{5}$ _____   **27.** $\frac{1}{2}$ _____   **28.** $\frac{7}{25}$ _____

29. $\frac{11}{50}$ _____   **30.** $\frac{1}{10}$ _____   **31.** $\frac{17}{20}$ _____   **32.** $\frac{3}{5}$ _____

33. 0.75 _____   **34.** 0.2 _____   **35.** 0.88 _____   **36.** 0.03 _____

37. 0.16 _____   **38.** 0.99 _____   **39.** 0.85 _____   **40.** 0.4 _____

Algebra Find each missing number.

41. $29\% = \frac{s}{100}$ **42.** $80\% = \frac{w}{5}$ **43.** $44\% = \frac{c}{25}$ **44.** $90\% = \frac{a}{10}$

$s = $ _____ $w = $ _____ $c = $ _____ $a = $ _____

Problem Solving
Solve.

45. A basketball team won 0.8 of its games. What percent of its games did the team win?

46. A basketball player made 23 out of 25 free throws. What percent of his free throws did the player make?

More About Percents

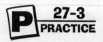

Write each percent as a decimal and as a mixed number in simplest form, or as a whole number.

1. 450% _____ **2.** 225% _____ **3.** 300% _____

4. 140% _____ **5.** 590% _____ **6.** 420% _____

7. 260% _____ **8.** 950% _____ **9.** 175% _____

10. 525% _____ **11.** 280% _____ **12.** 340% _____

13. 800% _____ **14.** 110% _____ **15.** 650% _____

16. 775% _____ **17.** 365% _____ **18.** 980% _____

19. 825% _____ **20.** 390% _____ **21.** 770% _____

22. 430% _____ **23.** 655% _____ **24.** 900% _____

Write each percent as a decimal and as a fraction in simplest form.

25. 37.5% _____ **26.** 49.6% _____ **27.** 12.5% _____

28. 6.5% _____ **29.** 45% _____ **30.** 31.5% _____

31. 35.5% _____ **32.** 29.6% _____ **33.** 57% _____

34. 43.5% _____ **35.** 87.5% _____ **36.** 43.2% _____

37. 14.5% _____ **38.** 38.4% _____ **39.** 64% _____

40. 24.5% _____ **41.** 20.2% _____ **42.** 94.6% _____

43. 71.3% _____ **44.** 78.5% _____ **45.** 82.8% _____

Problem Solving
Solve.

46. A softball player got hits in 37.5% of her at bats. Write a decimal to represent her batting average.

47. This year a baseball team increased its wins by 130% over last year. Write a decimal to show how many times more wins the team had this year than last.

_____ _____

Name _____

Problem Solving: Skill
Represent Numbers

Explain which numbers you would compare to solve each problem.
Then solve the problem.

1. Monica plays forward on her soccer team. Last year, 0.30 of her shots
   scored goals. This year, she made 16 goals out of 40. Did Monica
   improve her record this year? Explain.

   _____

   _____

2. Brian plays tournament table tennis. Last year, he won 72 percent of his
   games. This year, he has won 15 of his 20 games. Has Brian improved
   his record? Explain.

   _____

   _____

3. Jessica swims on a swim team. Last year, she placed first 12 times out
   of 20 in the breaststroke. This year, she has placed first 55 percent of
   the time. Was Jessica's record of winning better last year or this year?
   Explain.

   _____

   _____

4. Anja competes in chess tournaments. Last year, she won 12 out of 15
   games. This year, she has won 68% of her games. Was her record better
   this year or last year? Explain.

   _____

   _____

5. Elisa's family gets a pool table. Last month, 0.12 of the shots Elisa tried
   went in. This month, she has made 120 out of 500 shots. Is Elisa's game
   improving? Explain.

   _____

   _____

6. Peter's class takes timed division tests. Last month, Peter completed 66 percent of
   the problems correctly. This month, he has completed 60 out of 80 problems
   correctly. Has Peter improved his score? Explain.

   _____

   _____

Percent of a Number

Find the percent of each number.

1. 25% of 48 _____   **2.** 30% of 50 _____   **3.** 10% of 50 _____

4. 45% of 40 _____   **5.** 50% of 64 _____   **6.** 20% of 85 _____

7. 40% of 60 _____   **8.** 95% of 80 _____   **9.** 65% of 60 _____

10. 10% of $12.00 _____   **11.** 60% of $4.00 _____   **12.** 35% of $20.00 _____

13. 75% of $6.00 _____   **14.** 30% of $15.00 _____   **15.** 80% of $10.00 _____

16. 15% of $30.00 _____   **17.** 85% of $16.00 _____   **18.** 5% of $30.00 _____

19. 120% of 50 _____   **20.** 150% of 64 _____   **21.** 125% of 60 _____

22. 190% of 70 _____   **23.** 130% of 60 _____   **24.** 225% of 40 _____

25. 140% of $8.00 _____   **26.** 120% of $7.00 _____   **27.** 180% of $5.00_____

28. 225% of 84 _____   **29.** 55% of $7.00 _____   **30.** 150% of $15.00_____

Algebra Find each missing number.

31. ☐% of 90 = 9   **32.** 20% of ☐ = 5

33. ☐% of 40 = 10   **34.** 10% of ☐ = 7

35. ☐% of 60 = 12   **36.** ☐% of 80 = 40

Problem Solving
Solve.

37. Students pay 25% of the adult ticket price to attend ball games in town. An adult ticket costs $3.00. How much does a student ticket cost?

38. A football team wins 80% of the 10 games it played. A basketball team wins 45% of 20 games. Which team has won more games? Explain.

_____   _____

_____

Use with Grade 5, Chapter 28, Lesson 1, pages 648–651.

Percent That One Number Is of Another

Find the percent each number is of another. Round to the nearest whole percent, if necessary.

1. What percent of 60 is 48? _____

2. 24 is what percent of 40? _____

3. What percent of 32 is 16? _____

4. 15 is what percent of 20? _____

5. 33 is what % of 60? _____

6. What percent of 96 is 64? _____

7. 24 is what percent of 80? _____

8. What percent of 65 is 52? _____

9. 21 is what percent of 60? _____

10. 45 is what percent of 81? _____

11. What percent of 18 is 6? _____

12. What percent of 85 is 13? _____

13. 60 is what percent of 40? _____

14. 100 is what percent of 50? _____

15. What percent of 60 is 72? _____

16. 50 is what percent of 40? _____

17. What percent of 80 is 200? _____

18. What percent of 30 is 48? _____

19. 35 is what percent of 20? _____

20. What percent of 50 is 90? _____

21. 56 is what percent of 40? _____

22. What percent of 36 is 81? _____

23. What percent of 70 is 77? _____

24. 75 is what percent of 50? _____

Algebra Find each missing number.

25. _____ is 30% of 50.

26. _____ is 20% of 65.

27. _____ is 10% of 90.

28. _____ is 80% of 40.

Problem Solving
Solve.

29. A football team wins 12 of the 16 games it played. What percent of the games did it win?

30. Last year a football team won 10 games. This year it won 15 games. What percent of the games won last year did the team win this year?

_____

_____

Problem Solving: Strategy
Use Logical Reasoning

Use a Venn diagram to solve each problem.

1. Of 26 people surveyed, 19 said they go to basketball games and 12 said they go to football games. Five of the people said they go to both. How many people said they go to basketball games, but not to football games?

_____

2. Music Of 40 teachers surveyed, 34 said they listen to classical music and 17 said they listen to opera. Eleven of the teachers said they listen to both classical music and opera. How many teachers listen to classical music, but not to opera?

_____

3. Of 24 students surveyed, 17 students said they liked board games and 12 said they like card games. Five students said they liked both. How many students said they like board games, but not card games?

_____

4. Health Of the 50 people surveyed at a recreation center, 32 said they used the basketball courts and 24 said they used the racquetball courts. Six of the people said they used both courts. How many people said they use the racquetball courts, but not the basketball courts?

_____

Mixed Strategy Review

Solve. Use any strategy.

5. Nathan wants to buy trading cards. Superstar packages cost $3.23 each and mixed packages cost $1.78 each. Nathan buys 7 packages and spends a total of $15.36. How many of each type of package did he buy?

_____

Strategy: _____

6. An after-school club is building a clubhouse that is 8 feet by 6 feet. They are also including a trampoline with a radius of 4 feet. What is the total area of the clubhouse and the trampoline, to the nearest square foot?

_____

Strategy: _____

7. A band is performing on a rectangular stage that is 36 feet by 24 feet. The manager wants to set up lights every 4 feet around the stage, including the corners. How many lights will he need?

_____

Strategy: _____

8. Write a problem that you could use logical reasoning to solve. Share it with others.

_____

_____

_____

Circle Graphs

Use data from the circle graph for problems 1–5.

1. List the activities from favorite to least favorite.

_____

2. What fraction of the total votes went to in-line skating?

_____

3. If 200 people were surveyed, how many people said jumping rope was the favorite activity?

_____

4. If 140 people were surveyed, how many people said basketball?

_____

Favorite After-School Activity

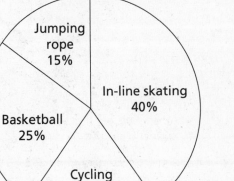

5. Write a statement that compares the results of the data.

_____

Use data from the table for problems 6–8.

6. List the number of degrees you would use to make each part of a circle graph to show the data.

Baseball: _____

Basketball: _____

Football: _____

Soccer: _____

7. Make a circle graph at the right to show the data.

8. If 120 people were surveyed, how many named baseball as their favorite spectator sport?

_____

Sport	Percent of Total Responses
Baseball	35%
Basketball	30%
Football	15%
Soccer	20%

Favorite Spectator Sport

Summer Skills Refresher

Summer Skills

Stadium Facts

The Sports Stadium is home to several teams. The table below contains some facts about the stadium.

> The Sports Stadium is a 70,000 seat stadium.
> The stadium contains: 30 private suites
> Press boxes to accommodate
> 250 writers
> Two 9,000 square foot locker rooms
> 50 concession stands and restrooms
> Its original cost in 1936 was $115,000.

1. Write all of the numbers from the table in word form:

   70,000: _____

   30: _____

   250: _____

2. The original stadium contained only 10,000 seats. How many more seats were added since 1936?

   _____

3. Suppose it costs $25 per ticket to enter the stadium. If a football game was sold out, how much money would the stadium receive from ticket sales?

   _____

4. The seating capacity for a baseball game in the stadium is only 5,500 people. How many more people can attend a football game in the stadium than a baseball game?

   _____

Answers: 1. seventy thousand; thirty; two hundred fifty; 2. 60,000 seats; 3. $1,750,000; 4. 64,500 people

Land Facts

Western forests generate $8 billion in manufactured products annually; they provide jobs for 115,000 people who are paid almost $1.25 billion. Private landowners own 47% of the forest, various companies own 35% of the land, and the government owns 18%.

5. If private landowners own 7 million acres, rounded to the nearest hundredth, how many million acres are in western forests?

_____

6. Use the answer from problem 5 to find how many million acres the government owns, rounded to the nearest hundredth.

_____

7. Write $1.25 billion in standard form.

_____

8. Suppose the private landowners generated 47% of the annual products manufactured. How much money did they generate?

_____

Answers: 5. 14.89 million acres; 6. 2.68 million acres; 7. $1,250,000,000; 8. $3.76 billion

Summer Skills

What I Did Last Summer

Sarah drew a map of Florida on a rectangular piece of poster board for her presentation describing what she did during summer vacation. During her vacation, she drove from her home in Tallahassee to visit her grandparents in Jacksonville. They all drove together to Orlando.

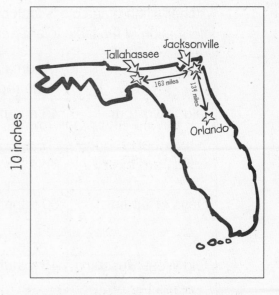

1. Look at the map Sarah drew. If she and her family drove from their home to her grandparents' house in 2.5 hours, about how fast did they drive?

_____

2. At one rest stop during the trip, Sarah's father put $16.75 in gasoline into their car. He paid with a $20 bill. How much change did Sarah's father receive?

_____

3. Suppose the length of the poster board Sarah used was 10 inches. Estimate the width of the poster board.

_____

4. Sarah wanted to put a border of ribbon around the edge of her poster. Use your estimate in problem 3 to find the perimeter of the poster board. About how much ribbon will she need?

_____

Answers: 1. 65.2 miles per hour; 2. $3.25; 3. 5 inches; 4. 30 inches

Laredo

Mitchell's fifth-grade class took a class trip to Laredo. Use the information in the table to answer problems 5–7.

	Approximate time required	Fee
Old Town Tour	45 minutes	$8 per person Children under 6 free
River Cruise	90 minutes	$8 per person Children under 6 free
Days of Spain	60 minutes	$6 per adult $3 per child age 6–18 Children under 6 free
Old West Museum	45 minutes	$7.95 per adult $4.75 per child age 5–18 Children under 5 free

5. Mitchell's class began the Old Town Tour at 10:05 A.M. About what time should they expect to finish their tour?

_____

6. Including himself, there are 15 students in Mitchell's class. They were chaperoned by 5 adults. They paid a group fee of $100 for the admission to the Old Town Tour. How much money did they save by paying as a group?

_____

7. Some of Mitchell's classmates decided to visit Days of Spain. The others decided to visit the Old West Museum. Both groups need to finish their tours at 12:30 P.M. to have enough time to make it to their bus. What is the latest time the students can start the tours of Days of Spain and the Old West Museum so they are outside at 12:30 P.M.?

_____

Answers: 5. 10:50 A.M. 6. $60. 7. Days of Spain: 10:45 A.M.; Old West Museum: 11:45 A.M.

Summer Skills

Art Project

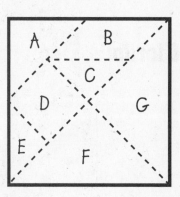

Wayne created the sun catcher on the right for his elementary school art fair. He used a variety of shapes to create his pattern.

1. Name each polygon in the pattern.

   Shape A: _____   Shape B: _____

   Shape C: _____   Shape D: _____

   Shape E: _____   Shape F: _____

   Shape G: _____

2. Shape F and Shape G are congruent. Describe how you could transform Shape F to match Shape G.

   _____

3. Name all of the shapes that are congruent. Name all of the shapes that are similar.

   Congruent shapes: _____

   _____

   Similar shapes: _____

   _____

4. Wayne colored all of the triangles blue and all of the quadrilaterals red. He outlines all of the squares in green. Color the pattern the same way Wayne did.

Answers: **1.** A triangle; B parallelogram; C triangle; D square; E triangle; F triangle; G triangle; **2.** reflection over side that the two triangles share, or rotate if 90 degrees about the top; **3.** C is congruent to E; F is congruent to G; A, C, E, F, and G are similar figures; **4.** check students' coloring

Florida's Flag

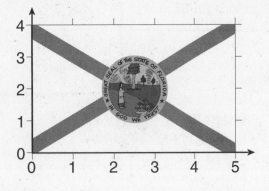

To help her copy the state flag onto paper, Carla used grid paper. She graphed the bottom left corner of the flag at (0, 0).

5. Name the missing coordinates of the corners of Carla's drawing?

_____

6. How do you know that Carla drew the flag as a rectangle?

_____

7. What is the measure of the angle with its corner at (5, 0)?

_____

8. Describe the shapes that make up the state flag of Florida.

_____

Answers: 5. (0,0) (5, 0) (0, 4); 6. opposite sides are congruent; 7. 90°; 8. triangles, circle, rectangle

Summer Skills

Computer Needs

The Computer Recycling Program repairs and upgrades donated computer equipment and then gives them away. Suppose in January they received exactly 5 donated computers from one company. Suppose 3 other companies donate an equal number of computers.

1. Let c represent the number of computers donated by each of the 3 companies. Write an expression to represent the number of computers donated.

   _____

2. Let t represent the total number of computers donated in January. Write an equation using c and t.

   _____

3. Complete the function table to represent the total number of computers donated.

c	0	1	2	3	4
t					

4. Plot the ordered pairs on a graph. Connect the points with a line.

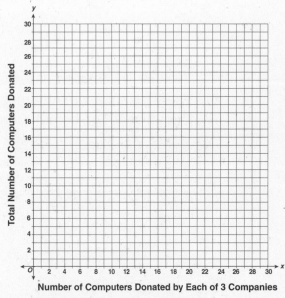

Total Number of Computers Donated

Number of Computers Donated by Each of 3 Companies

Answers: 1. 5 + 3c; 2. t = 5 + 3c; 3. 5, 8, 11, 14, 17; 4. check graph

Tropical Gardens Park

Kiera, her 2 brothers, 1 sister, and her mother and father decided to spend the day at Tropical Gardens Park.

5. Her family spent $28.50 on lunch. They each had a hot dog, a bag of chips, and a soda. A bag of chips costs $1.25. A soda costs $1.75. Let h represent the cost of each hot dog. Write an equation that you can use to find the value of h.

_____

6. Use your equation from problem 1 to find the cost of each hot dog.

_____

7. The safari is an off-road tour of African animals. Each tour has a capacity of 20 people. There are 80 people signed up for the safari before Kiera's family. Let x represent the number of full tours that will go before Kiera's family. Write an equation that you can use to find the value of x.

_____

8. Use your equation from problem 7 to find the number of full tours that will go before Kiera's family.

_____

Answers: **5.** $28.5 = 18 + 6h$ or $6(1.25 + 1.75 + h) = 28.5$; **6.** $1.75; **7.** $20x = 80$; **8.** 4 tours

Summer Skills

Fun Fishing

The Stein family spent a week camping and fishing off the coast of Maine. The pictograph below shows the number of fish they caught each day during their vacation.

Number of Fish Caught

1. Look at the pictograph. Let each picture represent 2 fish. How many fish did the Stein family catch on Thursday?

   _____

2. How many fish did the Stein family catch all together?

   _____

3. What is the mean number of fish the Stein family caught each day during their vacation?

   _____

4. What is the range in the number of fish the Stein family caught each day during the week of their vacation?

   _____

Answers: 1. 2 fish; 2. 28 fish; 3. 4 fish; 4. 8 fish

Survey

Jacob and Hannah wanted to survey people in their town on the number of times people swam at the local pool in the last month. They asked every tenth person that walked into the local supermarket.

5. Name the population and sample that Jacob and Hannah used.

_____

6. The survey produced the following results:
   1–2 times (10)
   3–4 times (12)
   5–6 times (15)
   7–8 times (25)
   9–10 times (23)
   More than 10 times (31)

   Use the axes to the right to draw a histogram to display the results of the survey. Label your survey results.

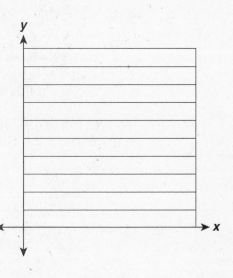

7. How many people swam 7–8 times last month?

_____

8. How many more people swam 5–6 times last month than 3–4 times?

_____